Process Development Division

Presentations at the 2008 AIChE Annual Meeting

Philadelphia, Pennsylvania
16 - 21 November 2008

ISBN: 978-1-61567-226-4

Printed from e-media with permission by:

Curran Associates, Inc.
57 Morehouse Lane
Red Hook, NY 12571

Some format issues inherent in the e-media version may also appear in this print version.

Printed by Curran Associates, Inc. (2009)

For permission requests, please contact AIChE
at the address below.

AIChE
3 Park Avenue
New York, NY 10016-5991

Phone: (203) 702-7660
Fax: (203) 775-5177

www.aiche.org

Additional copies of this publication are available from:

Curran Associates, Inc.
57 Morehouse Lane
Red Hook, NY 12571 USA
Phone: 845-758-0400
Fax: 845-758-2634
Email: curran@proceedings.com
Web: www.proceedings.com

TABLE OF CONTENTS

25c Acrylic Monomer Process Technology Evolution from Reppe to Renewables

Timothy Donnelly[1], Thomas S. Hackman[1], and Dan A. Bors[2]. (1) Primary Materials, Rohm and Haas Texas Inc, 6519 State Highway 225, Deer Park, TX 77536, (2) Primary Materials, Rohm and Haas Company, Spring House Technical Center, 727 Norristown Road, Spring House, PA 19477

This talk surveys the evolution of acrylic monomer technology from its origins in the early 20th century through the newest processes being developed today based on industrial biotechnology. First-generation acrylate technology was based on the Reppe process, using acetylene as a key intermediate and a nickel carbonyl catalyst to enable monomer synthesis. In the 1970's, environmental and operational challenges drove development of second-generation technology based on the two-stage conversion of propylene to acrylic acid over mixed metal oxide catalysts. As that technology enters its fourth decade, research focuses on new approaches that are both economical and sustainable. Acrylate manufacturers likely face a choice between continuing to exploit petroleum-based sources, for example via the partial oxidation of propane, or converting to new approaches using renewable feedstocks. Each option raises a series of technical, operational and ethical questions, which will be discussed in this presentation.

25d Strategies and Tactics Employed in the Development of Zoloft/sertraline Commercial Process

George Quallich, Research and Development, Pfizer Inc, Eastern Point Rd, Groton, CT 06340

Sertraline represented Pfizer's first product wherein the stereocenters were not derived from natural sources, as was the case with Zithromax and Beta-lactam products. Discovery chemists employed chromatography and classical resolution to secure the active agent's two stereocenters. Investigations to identify a commercial route included a number of synthetic approaches including enantioselective ketone reduction, SN2 displacement of an activated alcohol to transfer chirality, and continuous chromatography. The challenges of adopting new technology in the pharmaceutical industry are discussed in the context of Sertraline's development. The successful outcome employed continuous chromatography, and the potential for this technology to shift the development paradigm for any development candidate or commercial product is described.

All Hydroprocessing Route to High Quality Lubricant Base Oil Manufacture Using Chevron ISODEWAXING Technology

Guan-Dao Lei[1], *Arthur Dahlberg*[1] and *Kamala Krishna*[2]

(1)Chevron Products Company, 100 Chevron Way, Richmond, CA 94801,

(2)Chevron Energy Technology Company, 100 Chevron Way, Richmond, CA 94801

Extended Abstract

Introduction

Chevron has made significant, long-term commitments to the production of high quality petroleum products. This is especially true for lubricant base oils, where Chevron is a world leader in the production of premium base oils made exclusively by an all-hydroprocessing process technology route.

Early Base Oil Processing

Base oil technology has undergone many phases of evolution. In the 1920s, clay treating, acid treating and SO2 treating were used to remove some of the worst components in the petroleum base oil. These compounds were usually aromatics and highly polar compounds containing sulfur and nitrogen. By 1930s, solvent processing emerged as a viable technology for improving base oil performance using a fairly safe, recyclable solvent. Most base oil producers still use this process. Solvent refined base oils are commonly called Group I base oils which have higher aromatics and sulfur. Table I shows the API Base Oil Categories.

Table 1. API Base Oil Categories

Group	Sulfur, Wt %		Saturates	V.I.
I	>0.03	and/or	<90	80-119
II	□0.03	and	□90	80-119
III	□0.03	and	□90	≥120
IV	All Polyalphaolefins (PAOs)			
V	All Stocks Not Included in Groups I-IV (Pale Oils and Non-PAO Synthetics)			

Solvent processing has evolved over time but the basic processing steps remain the same. Aromatics are removed by solvent extraction in the first step, and wax is removed from the oil by chilling and precipitation in the presence of a different solvent in the second step. Finished oil quality improved mainly due to the addition of additives. For several decades, the lubricants industry continued to rely heavily on additive technology to improve the performance of finished oils. This also imposed limitations on the crude source of the starting material.

Lubes Hydroprocessing

Hydrotreating was developed in the 1950s and first used in base oil manufacturing in the 1960s as an additional "cleanup" step added to the end of a conventional solvent refining process. This process removed some of the nitrogen and sulfur containing molecules but was not severe enough to remove a significant amount of aromatic molecules. This made a small improvement in base oil quality until all-hydroprocessing technology was available.

Recent lube hydroprocessing include hydrocracking and catalytic dewaxing processes. Hydrocracking is a more severe form of hydroprocessing. Hydrocracking reshaped feed molecules by converting low Viscosity Index (VI) components to high VI base oil material and also by cracking low VI components to lower boiling range materials. A great majority of sulfur, nitrogen, and aromatics are also removed. In 1969, the first hydrocracker for base oil manufacturing was commercialized in Idemitsu Kosan Company's Chiba Refinery (a Chevron Licensee). This could replace the solvent extraction step for VI upgrade, or use in conjunction with solvent extraction to result in higher waxy base oil yield.

Catalytic dewaxing was developed in the 1970s to remove wax from oil. Catalytic dewaxing was a desirable alternative to solvent dewaxing because it removed n-paraffins and waxy side chains from other molecules by catalytically cracking them into light hydrocarbons. Catalytic dewaxing was used in place of solvent dewaxing, but still coupled it with solvent extraction to manufacture conventional base oils.

Chevron's lube hydroprocessing experience extends back to 1984. Chevron was the first to combine catalytic dewaxing with hydrocracking and hyrofinishing in their Richmond Lube Oil Plant (RLOP) in California (Figure 1). The hydrofinishing step was added to the lube hydroprocess to completely saturate aromatics to obtain excellent oxidation and thermal stability. This results in superior color and better response to additive in the finished lubes. In 1984, RLOP was started up to produce premium base oils from crude oils that are deficient in high VI components. These poor-lube-quality crudes, such as Alaskan North Slope and heavy California San Joaquin Valley, were readily available but primarily used at that time for fuels production. This was the first commercial demonstration of an all-hydroprocessing route for lube base oil manufacturing.

Figure 1. Chevron Richmond Lube Oil Plant (RLOP)

RLOP produced light, medium, and heavy neutral base oils in two parallel hydroprocessing trains. The Light Train, which made light and medium neutral oils, consisted of a hydrocracker followed by a catalytic dewaxer. The Heavy Train, which made heavy neutral, consisted of a hydrocracker and a solvent dewaxer.

During the next several years, Chevron developed a selective, wax isomerization catalyst. The culmination of the development effort was its commercialization in RLOP in 1993. The new process became known as ISODEWAXING and represented a breakthrough improvement in hydroprocessing technology. Instead of removing wax molecules (as in solvent dewaxing) or cracking them to light hydrocarbons (as in catalytic dewaxing), the ISODEWAXING® catalyst isomerizes the wax molecules into low pour point lubricating base oil. The ISODEWAXING® catalyst significantly increases lubricant base oil yield over competitive technologies (solvent dewaxing, catalytic dewaxing). It also can significantly improve the product VI due to the isomerization of wax (highest VI component) into the lube fraction.

When the ISODEWAXING process was commercialized in 1993, Chevron shut down the solvent dewaxer and made all of the lubricant base oils in the ISODEWAXING unit (converting from catalytic dewaxing). In 2000, as part of an expansion project, a separate ISODEWAXING unit was added to the Heavy Train. RLOP now consists of two parallel all-hydroprocessing trains.

By using an all-hydroprocessing route (combining Chevron's ISOCRACKING, ISODEWAXING, and ISOFINISHING technologies), feedstocks with poor lubricating qualities are transformed and reshaped into higher quality base oils. VI, Pour point, and oxidation stability are controlled independently. All three steps convert undesirable molecules into desirable ones. Among the many benefits of this combination of processes is greater crude oil flexibility which allows lube plant to be less reliance on a narrow range of crude oils. In addition, the base oil performance is exceptionally favorable and substantially independent of crude source, unlike solvent-refined base oil.

Base Oil Market Change

The need for higher fuel economy and higher performance engines created a market for higher quality base oils. Lube base oils made by hydroprocessing technologies showed favorably differentiated performance which prompted the API to categorize base oils by composition (API Publication 1509) in 1993, as shown in Table 1. Group II base oils made from the all-hydroprocessing route are differentiated from Group I base oils because they contain significant lower levels of impurities. From a lubricant performance standpoint, improved purity means that the base oil and the additives in the finished product can last much longer. The base oil market has gradually shifted from Group I to Group II base oils due to the new automotive engine oil specifications that favor Group II quality. Today more than half of the paraffinic base oil market in North America has now converted to Group II quality. Many Group I plants have been converted or closed.

The all-hydroprocessing route can also produce higher VI Group III base oils. This is typically done by increasing the severity of the hydrocracking process. In modern base oil manufacturing, VI, pour point, volatility, and oxidation stability can be independently controlled. These modern Group III oils have greatly improved oxidation stability and low temperature performance than conventional Group I and Group II base oils. They substantially match existing levels of performance in finished lube applications already established by traditional synthetic oils such as PAOs. As well-designed Group III base oils become abundant in the market place, the performance gap between Group III and PAOs (Group IV) is closing.

Looking at the base oil market, we see high quality Group II and Group III base oils emerging in almost all regions. Group II /III base oils are appearing mainly because of simple economics. Refiners are not building new solvent plants (Group I) because new Group II/III plants are more economic in the long run. They also give refiners more flexibility with both crudes and products. The world wide market share of Group II/III base oils has grown from about 11% in 1995 to about 26% today. Availability of and advances in Chevron's all-hydroprocessing route using ISODEWAXING technology have created world wide premium base oil markets. This is truly a remarkable transformation for any refining segment in recent history.

25f Development of Shell Higher Olefin Process (SHOP)

Jonathan Worstell, Hoder, Shell Global Solutions US, 3333 Highway 6 South, Room CP-109, Houston, TX 77082

This paper outlines the historical development of SHOP, Shell's Higher Olefin Process. It discusses the chemical discoveries required and the chemical processing technology necessary to conceptualize SHOP.

74a Crossing the River Styx: Beginning a Career In the Chemical Processing Industry

Jonathan Worstell, Hoder, Shell Global Solutions US, 3333 Highway 6 South, Room CP-109, Houston, TX 77082

Senior engineering students at university are courted by companies wanting to hire them. Unfortunately, the honeymoon is all too short, as newly graduated engineers learn during the company "on-boarding" process. This talk presents the pitfalls of "on-boarding", what the company expects of newly graduated engineers, the necessity for career planning, and the characteristics required of them for successful career development. In other words, this talk presents the "ins and outs" of corporate survival, especially during the first year.

74b Engineers In Manufacturing: Turn In Your Books for a Hardhat and Coveralls

Shawna S. Berg, Process Engineering and Evaluations, Shell Global Solutions (US) Inc., Westhollow Technology Center, P. O. Box 1380, Houston, TX 77251

Tired of having your nose glued into a textbook--learning about the theories of chemical engineering? Want hands-on experience in how things really work? Become a manufacturing engineer. This session describes a day in the life of a manufacturing engineer including responsibilities, keys to success, and how a position in manufacturing can help your career.

74c A Day In the Life of a Process Evaluator

Shilpa Damle-Mogri, Process Engineering & Evaluations, Shell Global Solutions (US) Inc., Westhollow Technology Center, 3333 Highhway 6 South, Houston, TX 77082

Do you have a B.S. or PhD in Chemical Engineering and are afraid of spending your first assignment stuck in a lab wondering if your work will amount to something? Are you interested in a job that allows you to influence decision makers by (1) understanding technology & answering the question, "will it work?!", and (2) assessing a technology's economic, strategic, and social merits? Then a Process Evaluator may be the perfect job for you. Process Engineering Evaluators have the best of several worlds. Not only do they work along side the lab scale experimentalists, developing commercial scale processes from bench chemistry, they also evaluate the economics of the proposed processes, comparing novel & existing technologies, and identifying economically attractive windows. They are able to mesh technology know-how with economics to guide research and development through the pilot stages and beyond. As techno-economists, process evaluators in business development groups have the opportunity to asses business needs & market environment & align them with strategic technical initiatives. We are research and development with the power to influence the direction of the business. Please join me for "A Day in the Life of a Process Evaluator" to find out more.

74d The Endless Career Opportunities In the Chemical Processing Industry

Raj Sengupta, Shell Chemical LP, One Shell Plaza, 910 Louisiana, Houston, TX 77002-4916

The petrochemical industry requires chemical engineers to help run and maintain its plants, and troubleshoot the various problems that arise. Chemical engineers also work in R&D, where they help with developing new processes, and optimizing old ones. But did you know there are roles well beyond your typical "chemical engineering" roles in the industry that are both challenging and utilize your chemical engineering knowledge. These roles require understanding the fundamentals of chemical engineering, while utilizing your critical thinking ability developed as an engineer. This paper outlines the various opportunities that chemical engineers have in the industry such as Strategy, Supply, Technical Sales, Product Development, and Management which all require in-depth chemical engineering knowledge to succeed.

74e Slipping In the Back Door. A Not So Traditional Approach to Becoming a Chemical Engineer

Mike J. Doll, HODer, Shell Global Solutions, 3333 Highway 6 South, Houston, TX 77082

We know the drill. Graduate high school, go to college, earn a bachelors degree (or higher) in Chemical Engineering, get a job, and have a rewarding career. For a majority of engineers, this process is the norm. What about those that either don't have the means and motivation exiting high school or choose a different career direction initially? Is there some way to become a chemical engineer without going back to college, paying ones dues, and consuming additional, valuable time? There is! This paper discusses the author's journey down the non-traditional path to becoming a chemical engineer.

200a Improving Pharmaceutical Process Design: From Batch to Continuous Flow Reactor

Michelle K. Kassner[1], Reagan Charney[2], Pamela Pollet[2], Kent Richman[3], Charles A. Eckert[1], and Charles L. Liotta[2]. (1) School of Chemical and Biomolecular Engineering, Georgia Institute of Technology, Atlanta, GA 30332-0100, (2) School of Chemistry and Biochemistry, Georgia Institute of Technology, Atlanta, GA 30332-0400, (3) American Pacific Corporation, Las Vegas, NV 89109

Continuous flow reactors have been used in the chemical industries for decades. They are often preferable to batch systems due to enhanced reproducibility and scalability, improved yields, and better control over processing conditions. While the pharmaceutical industry has traditionally employed batch reactors, many manufacturing processes are beginning to incorporate continuous flow reactors. Low volume, high throughput continuous flow reactors are especially suited for reactions containing high exotherms and/or dangerous starting materials.

Here we consider the challenges faced when converting a traditional batch process to a continuous flow system for the production of a pharmaceutical intermediate. The reaction of interest consumes diazomethane, a hazardous material, and also produces a temperature-sensitive intermediate. We examine the optimization process for this reaction involving size scale, flow rate, concentration, and temperature. Also, the inherent safety advantages of a continuous flow system for this type of reaction are discussed.

PROCESS INTENSIFICATION IN CONTINUOUS BASE-CATALYZED BIODIESEL PRODUCTION

Joanna McFarlane, Joseph F. Birdwell Jr., Costas Tsouris, Hal L. Jennings
Oak Ridge National Laboratory, Oak Ridge, TN

Abstract

Recent expansion of biodiesel manufacture has resulted in increased interest among commercial enterprises to minimize the cost of feedstock materials and waste production and to maximize the efficiency of production. Oak Ridge National Laboratory has experience in developing process intensification methods for nuclear separations, and this paper will discuss how technologies developed for very different applications have been modified for continuous reaction/separation of biodiesel. In collaboration with an industrial partner, this work addresses the aspect of base-catalyzed biodiesel production that has previously limited it to a slow batch process. In particular, we have found that interfacial mass transfer and phase separation control the transesterification process and have developed a continuous two-phase reactor for online production of a methyl ester and glycerol. Enhancing the mass transfer has additional benefits such as being able to use an alcohol-to-oil phase ratio closer to stoichiometric than in conventional processing, providing opportunities to minimize the amount of solvent that has to be recycled and reduce post-processing clean up. Various technical issues associated with the application of process intensification technology will be discussed, including scale-up from the laboratory to a pilot-scale undertaking.

Introduction

Process intensification is an engineering concept that gained attention through investigations in the 1970s at the University of Newcastle [1]. The goal was to achieve simultaneous reduction in capital cost of a chemical plant through reduced size and inventory, as well operating costs through reduced energy consumption and feedstock required per unit mass of the product. Since the idea was put forward, it has been applied to a number of chemical processes, perhaps most notably the Eastman technology for the esterification of methanol to form methyl acetate, which combines reaction and distillation [2]. More recently, process intensification has been examined as a result of imperatives in chemical safety and safeguards. For instance, the Bhopal incident would not have affected as many people had the inventory of methyl isocyanate been lower. Finally, in the era of dwindling supplies of easily recoverable oil, sustainability depends on dependable supply of both feedstock and energy. Methods of process intensification naturally enhance both environmental and economic sustainability of the chemical industry. Process intensification has been reviewed by Tsouris and Porcelli [3,4], who cite examples including: reactive distillation, static mixers, "HiGee" reactor/separators such as spinning disc [5] and centrifugal contactors [6], fractal technologies to control fluid flow [7], heat-exchanger reactors, microreactors, and process equipment that is enhanced by external fields [8] – electric, sound, magnetic, gravity and microwaves.

Centrifugal phase contact and separation is an example of an intensified technique that enhances mass transfer at high throughput and minimizes inventory of solvents. Such methods are well suited for applications in treatment of nuclear waste [9] and nuclear fuel recycling [10]. Contactors for such uses are available commercially and have been tested in processing of actual radioactive feeds [11]. Contactor technology has also been tested in the area of oil recovery, enhancing the separation of produced water from the lighter hydrocarbon phase [12]. In all of these solvent extraction applications, the process has been optimized to maximize mixing followed by phase separation with minimal residence time, necessary in nuclear applications to minimize solvent degradation from radiolysis. The work that is discussed in this paper applies the well-known centrifugal contactor/separator technology to a different system that is well suited for process intensification – biodiesel production.

Biodiesel, a mixture of methyl esters, is made commercially from the transesterification of oil [13], often soy oil, Reaction 1. The kinetics of the transesterification process are rapid; however, multiphase separations after the synthesis of the fuel can be problematic and so the process is typically run in batch mode. The biodiesel fuel and the glycerol product take several hours to separate. In addition, to push yields to completion, an excess of methoxide catalyst is typically used, which has to be removed from both the biodiesel and the glycerol phase after reaction. Washing steps are often employed to remove free fatty acids, which can lead to undesirable saponification. Standards for biodiesel purity are based on the removal of contaminants either before the oil feedstock is esterified or separation of unwanted byproducts [14].

$$C_3H_5(CO_2R^1)(CO_2R^2)(CO_2R^3) + 3CH_3OH \rightarrow CH_3O_2R^1 + CH_3O_2R^2 + CH_3O_2R^3 + C_3H_8O_3 \qquad (1)$$

Triglyceride + Methanol ⟶ Methyl Esters + 1,2,3-Propanetriol

Various methods have been examined to enhance either the pretreatment of biodiesel feedstocks or posttreatment of reaction products, including the use of cavitation reactor in the process intensification of the homogeneous acid catalysis of transesterification [15]. Centrifugal mixing has been applied to biodiesel production, using the contactor as a low throughput homogenizer, employing very low flow rates to increase residence times to tens of minutes [16]. In this study, we have combined the reaction of oil and methoxide with the online separation of biodiesel and glyercol into one processing step, using a modified centrifugal contactor. Two distinct phases enter the reactor (reagents) and two distinct phases leave the reactor/separator (products), demonstrating the application of process intensification to high-throughput biofuel production.

Experimental

Based-catalyzed biodiesel synthesis and simultaneous separation of methyl ester and glycerol products was carried out in a modified centrifugal contactor [17][1]. Reagents were food grade soy oil and potassium methoxide, formed from the reaction of KOH (EM Science) and methanol (EM Science 99.8%), at a phase ratio of 5:1 oil-to-methoxide. All reagents were used without further purification. Analysis of reaction products was by gas chromatography flame-ionization detection, GC FID, with a Hewlett Packard 5890 II GC. The analysis procedure followed ASTM D6854 for [18].

Before using the reactor/separator, the kinetics of the based-catalyzed transesterification reaction were studied by sampling a homogenized (IKA Labortechnik model T25) mixture of the oil and methoxide, 120 mL, contained in a 300 mL beaker, and held at 56 ± 4°C. Samples of two mL were

[1] Patent pending

taken after 15, 30, 60, 120, and 240 s, with t = 0 s corresponding to when the reaction started. After sampling, the samples were immediately contacted with 1 mL of 0.1 N HCl to arrest the transesterification reaction. Samples were analyzed by GC.

Once the required residence time was established, the reaction was carried out in a centrifugal contactor, modified to allow a sufficient contact time to permit the reaction to go to completion. The reagents, introduced through separate ports, were combined in a centrifugal contactor with the internal rotor spinning at 3600 rpm, Figure 1. Mixing to enhance mass transfer and reaction occurred, followed by separation of products according to their densities. Products were collected from the "light" side and "heavy" side outlet streams. The whole assembly was heated to 60°C in a temperature-controlled water bath, and the reagents were likewise preheated. Samples were taken after 1, 3, 6, and 10 min intervals and, as in the case of the kinetics experiments, were immediately contacted with a solution of 0.1N HCl. The samples were analyzed by GC to assess the extent of the reaction.

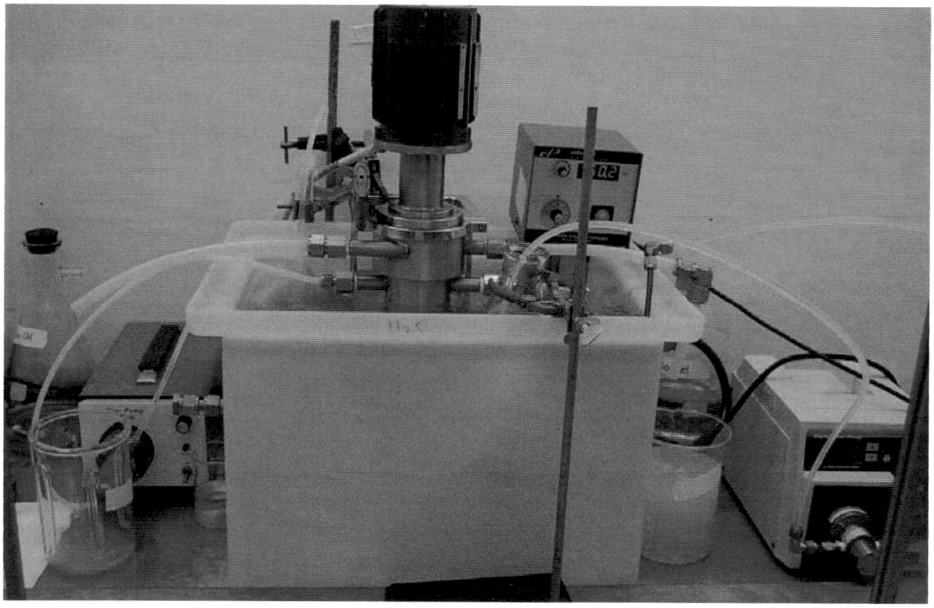

Figure 1: Picture of reactor/separator experimental apparatus in the water bath.

Results and Discussion

Results of the kinetics experiments are shown in Figure 2. The yield was calculated from the total GC signal for C16, C18, C20, C22, and C24, indicating the chain length of the hydrocarbon chain on the ester. GC signal included peaks from unsaturated as well as saturated methyl esters. The same data are presented in two different ways on the graph, the ratio of methyl ester-to-oil signal from the GC, and the second showing the same data converted to mole % biodiesel. There is a rapid increase in biodiesel fraction up to about 1 minute, and then the signal levels off. Although there appears to be a slight decrease in signal over the remaining data points, this is not statistically significant as the GC precision is about ±10%. The uncertainty in the abscissa is ±2s, during which grab samples were taken.

The analyses of product streams from the centrifugal contactor synthesis are shown in Figures 3 and 4. The bars show the fraction of unreacted oil versus methyl ester coming from the "light" and "heavy" fractions respectively. The flow of glycerol-rich dense phase was much slower than that of the biodiesel, and so during the initial stages of the experiment, methyl ester was found in both of the streams. After a minute the flows settled down, and the methyl ester and glycerol were well separated.

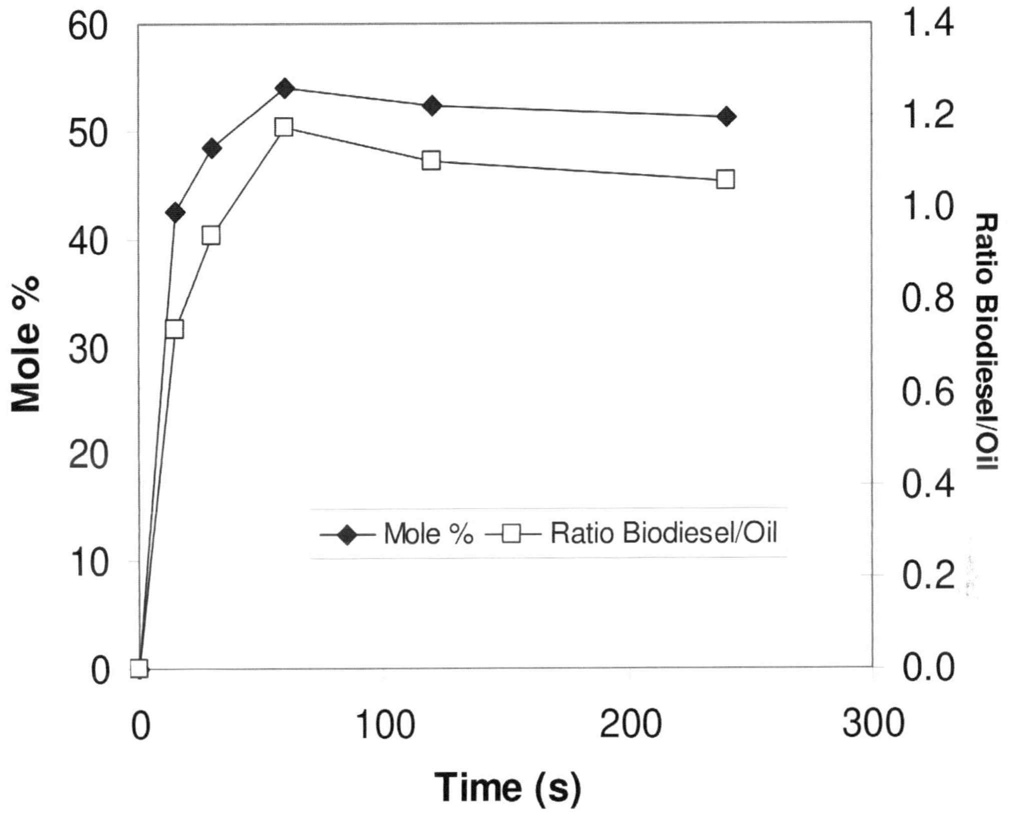

Figure 2: Mole % Biodiesel and Biodiesel-Oil Ratio as a Function of Time Determined from GC Analysis of Homogenized Samples.

Except at the beginning of the experiment, where mixing and flows were being established, no unreacted oil was present in the "light" side fraction. The data in Figure 4 show that over time, the fraction of methyl ester in the "heavy" side flow decreased dramatically as the flow of glycerol was established. However, because the system was configured to achieve the highest purity on the "light" side, some oil and methyl ester were entrained in the "heavy" side. The residual samples were taken well after the timed experiment was complete, and arose from the phases that were drained out of the bottom of the reactor. These samples showed that the centrifugation process was more effective than what could be achieved in a mixer settler relying on several hours of gravity settling for separation of the product and reagent phases. The effective separation in the contactor arose because of enhanced mass transfer from application of centrifugal forces equivalent to approximately 300 times that of gravity. The slight drop in signal at 6 and 10 minutes after the experiment started, as shown in Figure 3, is likely indicative of the precision of the GC method for methyl ester sampling, or about ±10%, as there was no deliberate variation in mixing speed or any other experimental parameter.

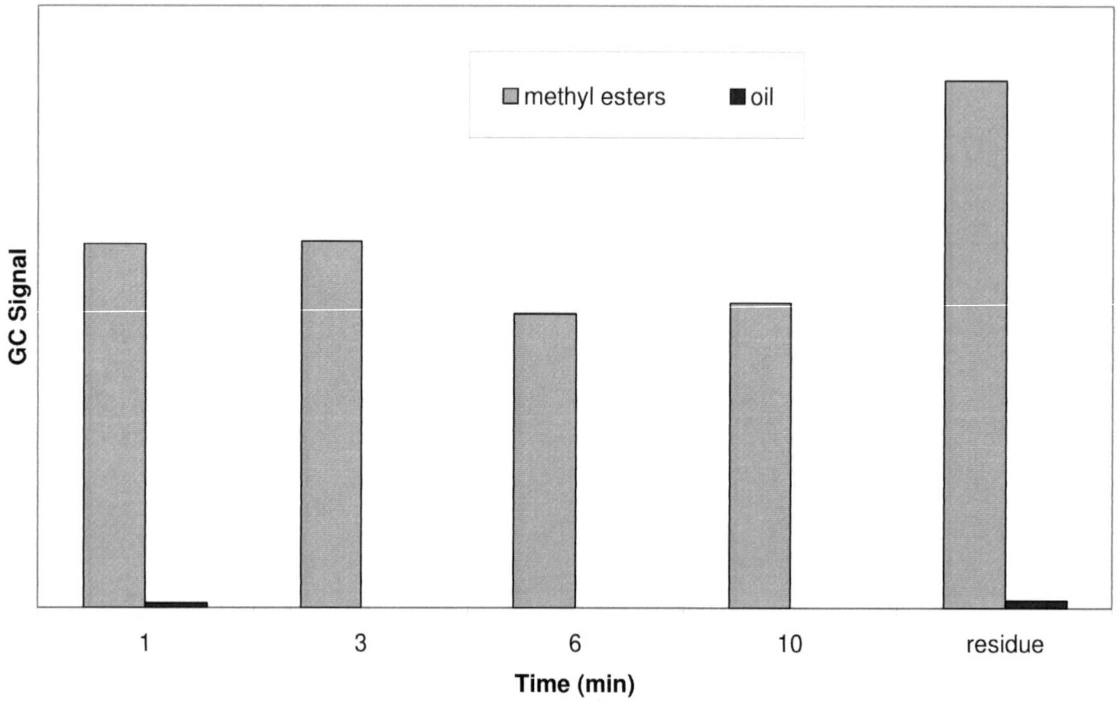

Figure 3: Continuous production of biodiesel showing less-dense flow from the reactor.

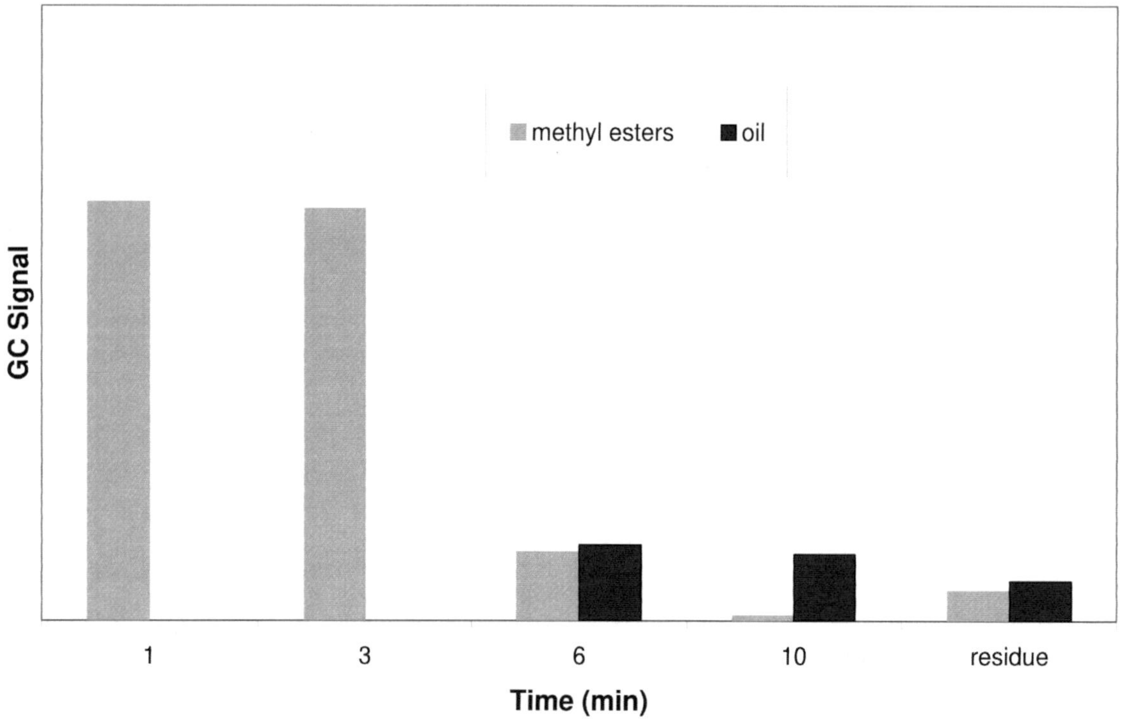

Figure 4: Continuous production of biodiesel showing more-dense flow from the reactor.

Figure 5 shows superimposed gas chromatograms for unreacted oil and the methyl ester product. The oil chromatogram, with a peak at 3.7 minutes, is from direct injection of the reagent oil before processing. Glycerol, expected to be seen at 8.4 minutes from injection through measurement of a calibration standard, was not observed in the chromatogram for the biodiesel product. Partially reacted mono- and diglycerides were also not observed. Although the system is complex, with a reaction involving two immiscible phases of reagents and two separate immiscible phases of products, these data indicate that the combination of reaction and separation is an effective means to synthesize biodiesel in a continuous manner.

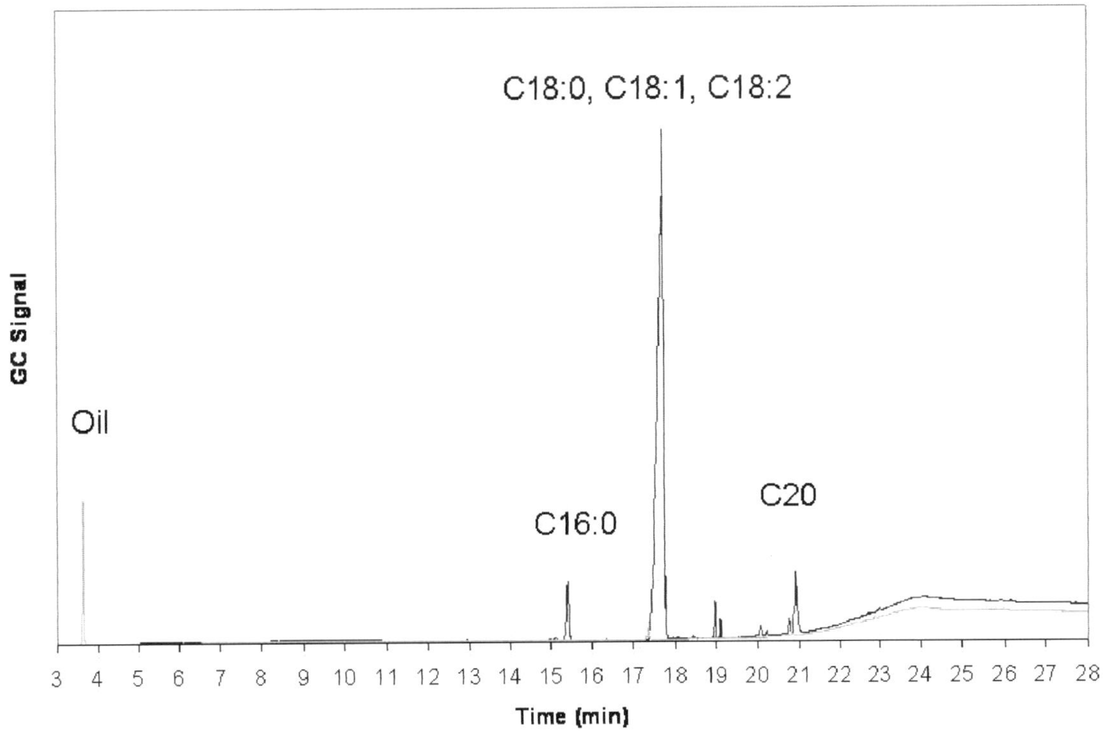

Figure 5: Gas Chromatogram of "light" side product from reactor/separator showing methyl ester peaks (black), superimposed with a chromatogram of unreacted oil (grey) taken at a separate time.

Conclusions

Simultaneous reaction/separation for the production of methyl esters from soy oil has been demonstrated in a centrifugal contactor, modified to achieve a residence time sufficient for the reaction to go to completion. Our approach is to operate the contactor under normal conditions, at several thousand rpm, but to engineer the hold up so that a high throughput can be obtained. Oak Ridge National Laboratory is now collaborating with a biodiesel manufacturer, Nu-Energie to scale-up the apparatus so that a pilot scale test can be done at the Nu-Energie facilities with feedstocks used in commercial production. It is anticipated that success with the pilot scale may lead to further applications in pre-and post-treatment in the biodiesel production process, allowing biodiesel production with less-than-optimal, but much less expensive, feedstocks.

Acknowledgments

This research was sponsored by the Laboratory Directed Research and Development Program of Oak Ridge National Laboratory (ORNL), managed by UT-Battelle, LLC for the U. S. Department of Energy. The authors are thankful to Ms. Denise Schuh for assisting in the experiments. The authors thank Nu-Energie, especially Mr. Brian Hullette and Mr. Joel Day, for their encouragement and feedback.

References

1. Stankiewicz, A. and J.A. Moulijn (2002), "Process Intensification," *Ind. Eng. Chem. Res.,* 41, 1920-1924.
2. Gogate, M.R., J.J. Spivey, J.R. Zoeller (1997), "Synthesis of methyl methacrylate by vapor phase condensation of formaldehyde with propionate derivatives," *Catal. Today*, 36(3), pp. 243-254.
3. Tsouris, Costas, and J.V. Porcelli (2003), Report on Topical Conference and Workshop, AIChE Spring National Meeting (April) (available at www.aiche.org/cwrt).
4. Tsouris, Costas, and J.V. Porcelli (2003), "Process Intensification – Has Its Time Finally Come?" *Chem. Eng. Prog.* 99(10), pp. 50-55.
5. Burns, J.R., C. Ramshaw, and R.J. Jachuck (2003), "Measurement of Liquid Film Thickness and the Determination of Spin-Up Radius on a Rotating Disc Using an Electrical Resistance Technique," *Chem. Eng. Sci.,* 58, pp. 2245-2253.
6. Birdwell, J.F. Jr., J. McFarlane, R.D. Hunt, H.M. Luo, and D.W. DePaoli (2006), "Separation of Ionic Liquid Dispersions in Centrifugal Solvent Extraction Contactors," *Sep. Sci. Technol.,* 41, pp. 2205-2223.
7. Kearney, M. (1999), "Control, of Fluid Dynamics with Engineered Fractals – Adsorption Process Applications," *Chem. Eng. Commun.*, 173, pp. 43-52.
8. Tsouris, C., C.T. Culbertson, D.W. DePaoli, S.C. Jacobson, V.F. De Almeida, J.M. Ramsay (2003), "Electrohydrodynamic Mixing in Microchannels," *AIChE J.*, 49, pp. 2181-2186.
9. Geeting, M.W., E.A. Brass, S.J. Brown and S.G. Campbell (2008), "Scale-up of Caustic-Side Solvent Extraction Process for Removal of Cesium at Savannah River Site," *Sep. Sci. Technol.*, 43, pp. 2786-2796.
10. Taylor, R.J., and I. May (2001), "Advances in Actinide and Technetium Kinetics for Applications in Process Flowsheet Modeling," *Sep. Sci. Technol.*, 36, pp. 1225-1240.
11. Leonard, R.A., J.; Ziegler, A.A., 1980. "Annular Centrifugal Contractors for Solvent-Extraction," *Sep. Sci. Technol.,* 15 (4), pp. 925-943.

12. Klasson, K.T., P.A. Taylor, J.F. Walker, S.A. Jones, R.L. Cummins, and S.A. Richardson (2005), "Modification of a Centrifugal Separator for in-Well Oil-Water Separation," *Sep. Sci. Technol.*, 40, 453-462.
13. Freedman, B., E.H. Pryde, and T.L. Mounts (1984), "Variables Affecting the Yields of Fatty Esters from Transesterified Vegetable Oils," *J. Amer. Oil Chem. Soc.*, 61(10), pp. 1638-1643.
14. ASTM Standard D6751-07b (2008), "Standard Specification for Biodiesel Fuel Blend Stock (B100) for Middle Distillate Fuels", ASTM International.
15. Keller, M., A., P.R. Gogate, and A.B. Pandit (2008), "Intensification of Esterification of Acids for Synthesis of Biodiesel Using Acoustic and Hydrodynamic Cavitation," *Ultrasonics Sonochemistry*, 15, pp. 188-194.
16. Kraai, G.N., F. van Zwol, B. Schuur, H.J. Heeres, and J.G. de Vries (2008), "Two-Phase (Bio)Catalytic Reactions in a Table-Top Centrifugal Contact Separator," *Angew. Chem.*, 47, 3905-3908.
17. Tsouris, C., J. McFarlane, J.F. Birdwell Jr., and H.L. Jennings (2008), "Intensive Production of Biodiesel via an Intensified Reactive/Extractive Process," in *Proceedings of the International Solvent Extraction Conference*, Tuscon, AZ, September15-19.
18. ASTM Method D6854 (2007) "Standard Test Method for the Determination of Free and Total Glycerin in B-100 Biodiesel Methyl Esters by Gas Chromatography," ASTM International.

200c High-Gravity Process Intensification Technology and Commercial Applications

Jian-Feng Chen and Lei Shao. Research Center of the Ministry of Education for High Gravity Engineering and Technology,, College of Chemical Engineering, Beijing University of Chemical Technology, Beijing, 100029, China

High-gravity technology is carried out in a rotating packed bed (RPB) and regarded as one of the important technologies for process intensification. The Research Center of the Ministry of Education for High Gravity Engineering and Technology in China has focused on the research of high-gravity technology for nearly two decades and made great progress in the application of this technology to process intensification. This proposal presents the application of high-gravity technology for the production of butyl rubber, petroleum sulfonate and overbased lubricating oil detergent with petroleum calcium sulfonate. The intensification of methylene diphenyl diisocyanate (MDI) production by high-gravity technology is also demonstrated.

200d Oscillatory Baffled Reactors as Bioreactors

Adam P. Harvey[1], C. N. Tröger[1], and S. Donegan[2]. (1) Ceam, Newcastle University, Merz Court, Newcastle University, Newcastle, United Kingdom, (2) Bioprocessing, CPI, Wilton Centre, Wilton, Middlesbrough, United Kingdom

Oscillatory baffled reactors (OBRs) are a form of continuous reactor, in which tubes fitted with orifice plate baffles have an oscillatory motion superimposed upon the net flow of the process fluid. The interaction of the baffles with the oscillatory motion of the fluid generates excellent mixing and enhanced transport rates, whilst maintaining a close approach to plug flow. Unlike conventional tubular reactors, where a minimum Reynolds number must be maintained, the tube-side mixing is independent of any net flow, potentially allowing long residence times to be achieved in a reactor of greatly reduced length-to-diameter ratio. This leads to a niche application in converting long reactions from batch to continuous processing.

An obvious class of long reactions is bioreactions /fermentations. As one major advantage of the OBR is very good uniform mixing, providing a good mass and oxygen transfer, it seems ideal for microbial purposes. The possibility of continuous fermentation is a second advantage as it will minimise the substrate and product inhibition. Lower and more uniform shear rates compared to a stirred tank reactor may be an additional advantage depending on the shear sensitivity of cultures such as mammalian, insect and plant cells, fungi and certain types of algae.

To investigate the possible advantages in this class of reactions, biopolymer, bioethanol and beer production have been investigated at various scales. The design modifications that have been required will be discussed. OBR processing has been shown to result in reduced residence times. Due to its plug flow RTD, the reactor also has the potential to allow various staged additions and parameter profiles along its length.

Web Page: pig.ncl.ac.uk

200e Solvent-Free Microwave Extraction of Essential Oil from Lavandin Super

Rafael B. Mato, Victor Martin, Alexander Navarrete, and Maria J. Cocero. Departamento de Ingenieria Quimica y Tecnologia del Medio Ambiente, Universidad de Valladolid, Facultad de Ciencias, 47011 Valladolid, Spain

Essential oils are traditionally obtained from plants by steam distillation. This is an effective method but demands a high energy consumption and a long processing time. Solvent-Free Microwave Extraction (SFME) is a new microwave based distillation process [1] which uses no additional added water or solvent. Preliminary studies [2] show SFME as a promising technique since energy and processing time are dramatically reduced, while oil quality remains unaffected, when compared to the traditional steam distillation process. This improvement in the performance of the process is based on the elimination of heat and mass transfer resistances which limit the oil extraction rate in steam distillation.

Scale-up of these preliminary experiences to a pilot-plant continuous process requires a deeper knowledge of the process. This knowledge must be translated into a mathematical model which represents the influence of process variables on the performance of the extraction process.

In this paper a phenomenological model of the process is developed and checked against batch experimental data. This model is developed from a previous one [3] obtained to describe essential oil steam distillation of lavandin super (Lavandula latifolia x angustifolia). Modifications are based on theoretical knowledge and experimental evidence.

References

[1] Lucchesi ME, Chemat F, Smadja J, "An original solvent free microwave extraction of essential oils from spices", Flavour and Fragrance Journal 19 (2) pp. 134-138 (2004)

[2] Lucchesi, ME, Chemat, F, Smadja, J, "Solvent-free microwave extraction of essential oil from aromatic herbs: comparison with conventional hydro-distillation", J Chromatogr. A 1043(2), 323-327 (2004)

[3] Cerpa MG, Mato RB, Cocero MJ, "Modeling steam distillation of essential oils: Application to lavandin super oil", AIChE J., 54(4), 909-917 (2008)

200f Influence of Different Mixtures of Catalyst and Adsorbent on the Smbr Performance: Reactive-Separation and Regeneration Regions

Viviana Silva and Alírio E. Rodrigues. Laboratory of Separation and Reaction Engineering, University of Porto - Faculty of Engineering, Rua Dr. Roberto Frias s/n, 4200-465 Porto, Porto, Portugal

Oxygenated compounds like acetals, esters and ketals are often produced by a reversible liquid phase reaction catalyzed by acids, where water is formed as subproduct. By means of chromatographic reactors, as the Simulated Moving Bed Reactor (SMBR), it is possible to integrate reaction and products separation into a single unit, displacing the equilibrium towards the products formation by removing at least one of them from the reaction medium. In addition to complete reactants conversion, operational and equipment costs are reduced. Usually, strongly acidic ion exchange resins, such as Amberlyst 15wet, are used both as catalyst as a selective adsorbent for water. However, water is strongly adsorbed and a large consumption of desorbent is needed in order to achieve high product purity and high productivity. Since the stationary phase and equipment costs are small with respect to the solvent recycle cost [1], it is desirable to choose an appropriate stationary phase for an SMBR that will provide the maximum productivity value but for a low desorbent consumption.

The aim of this work is to present the influence of the stationary phase on the SMBR performance by using mixtures of Amberlyst 15wet (a catalyst and selective adsorbent for water) with Smopex 101 fibres (catalyst), for the case of 1,1-dimethoxyethane production [2, 3]. By means of a complete model accounting for the different particularities of both catalyst and adsorbent (external mass-transfer resistance followed by reaction at the surface of Smopex 101 fibres; external and internal mass-transfer resistances for Amberlyst 15wet using a bi-LDF approach and adsorption and reaction at both particle and pores surface) the performance of the SMBR is studied. The concepts of reactive-separation region ($\gamma 2$-$\gamma 3$ plot) and regeneration region ($\gamma 1$-$\gamma 4$ plot) [4] are used to analyze the effect of the stationary phase composition in the productivity and desorbent consumption, respectively.

References:

[1] G. Ströhlein, Y. Assunção, N. Dube, A. Bardow, M. Mazzotti, M. Morbidelli, Chem. Eng. Sci., 61 (16), 5296-5306 (2006).

[2] C. Pereira, P. Gomes, G. Gandi, V. Silva, A. Rodrigues, Ind. Eng. Chem. Res. (ie070889t) (2008).

[3] G. Gandi, V. Silva, A. Rodrigues, Chem. Eng. Sci., 62 (7), 907-918 (2007).

[4] P. Sá Gomes, M. Minceva, A.E. Rodrigues, Adsorption, 12, 375-392 (2006).

Development of a new Synthesis Gas Production Catalyst and Process with CO_2 and H_2O Reforming

Tomoyuki Mikuriya, Fuyuki Yagi and Mitsunori Shimura

1. Introduction

Production of synthesis gas from natural gas and conversion of synthesis gas to fuels and chemicals would become increasingly attractive due to the viewpoint for energy resource diversification and raising the price of crude oil. And the technology of synthesis gas production by reforming of natural gas with CO_2 and H_2O is one of the effective solutions for the issues of energy resource and environmental preservation.

We have been investigating a new synthesis gas production catalyst and process with CO_2 and H_2O reforming of natural gas. In this process, synthesis gases of various molar ratios of hydrogen to carbon monoxide are produced with high energy efficiency and low amount of carbon dioxide emission. Because the proprietary reforming catalyst with high resistance against carbon deposition could allow carbon dioxide and steam reforming reaction under optimal operation conditions with low carbon dioxide and steam to feed gas ratio.

In this paper, the characteristics and performance of the catalyst are introduced. And also the process study results of GTL, oxo-alcohole and acetic acid production with our synthesis gas production process are represented.

2. Characteristics and performance of the catalyst

The activity and stability of group Ⅷ metal supported catalysts for CO_2 and H_2O reforming depend crucially on both the metal and the support. It has been suggested that noble metals are less active for carbon formation than Ni and the carbon deposition is suppressed when the metal is supported on a metal oxide with strong Lewis basicity.[1,2] Metal crystal structure and metal-support interaction may also be important to design the catalysts with the high resistance for carbon formation.[3] Based on these considerations, we have designed a new noble metal supported catalyst for CO_2 and H_2O reforming.

Fig.1 shows a result of EPMA(Electron Probe Micro Analyzer) analysis on our commercial size catalyst. The 4 peaks in left side figure indicate the concentration distribution of noble metal in the catalyst. Noble metal was loaded selectively on the outer surface of catalyst supports in egg-shell shape. As shown in this figure, metal content of inside is extremely lower than that of outside. It is considered the effectiveness factor of reforming catalyst is less than 0.1, therefore, metals on the outer surface can only contribute to reforming reaction. So in our catalyst high price noble metal is effectively utilized with consideration for characteristics of reforming reaction.

Fig.1 Result of EPMA Analysis of the Catalyst

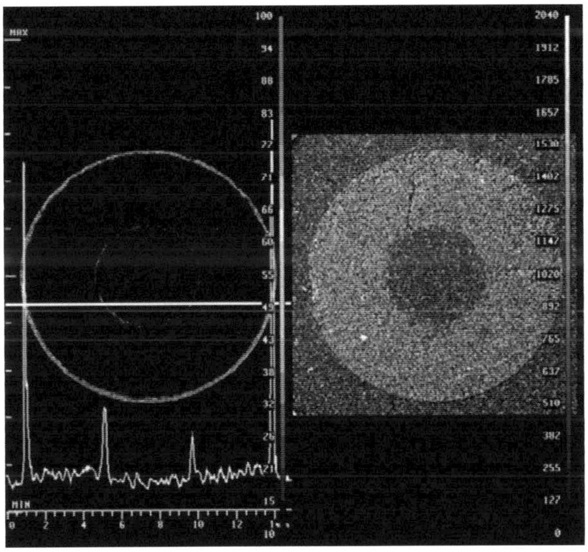

The pilot plant test of our CO2 and H2O reforming catalyst was conducted. The mono tube reformer whose size is inner diameter of 110mm and length of 12m was applied for syngas production with hydrogen to carbon monoxide ratio of 2.0. This tube size is almost same as the commercial reformer tube. Around one hundred litter of the reforming catalyst were charged in the reformer tube. Design temperature and pressure are 1025℃ and 2.4MPaG, respectively. Average syngas production rate is around 400Nm3/hr. Feed natural gas contains around 85% of methane and the rest are almost higher hydrocarbon than methane. This natural gas is reformed by the steam and carbon dioxide in the mono tube reformer without per-reforming in this pilot plant.

Fig.2 presents pilot plant test results. At the start up condition, low pressure and high steam to carbon and carbon dioxide to carbon ratio was taken. In the normal operation condition, the temperature of catalyst bed outlet was set around 900℃. Operation pressure was 1.9MpaG. And ratios of carbon dioxide to carbon and steam to carbon of the feed stream were controlled to 0.4 and 1.15

respectively. As shown in this figure, the stable operation for around 7,000 hours has been confirmed under the target reaction condition. After the operation, catalyst was discharged and analyzed. The amounts of the carbon deposition on the catalysts were less than 0.1wt% in the each part of the catalyst beds. From the results, the carbon-free operation has been demonstrated.

Fig.2 Pilot Plant Test Result of CO2 and H2O Reforming Catalyst

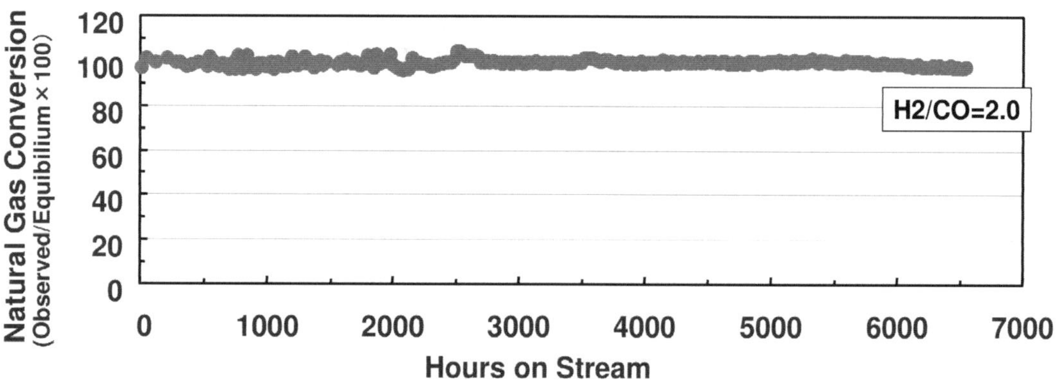

Temperature (Catalyst bed outlet) : 865～895℃, Pressure : 1.5～1.9MPaG
Feed Molar Ratio Hydrocarbon/CO_2/H_2O = 1.0/0.4～0.6/1.15～1.64

Fig.3 shows the results of catalyst performance test in the Lab. scale for production of syngas with ratio of 1.0. Typical reaction conditions are as follows; Temperature of 850 ℃ ,Pressure : 2.0MPaG, Feed Molar Ratio Hydrocarbon/CO2/H2O/H2 = 1.0/1.07/0.51/0.04. The 30cc of 3mm pellet catalyst were charged in the reactor. Long-run test for the production of syngas with low H2/CO ratio was attained around 1000hrs without catalyst deactivation.

Fig.3 Lab. Test Results the Catalyst for synthesis gas production of H2/CO=2.0

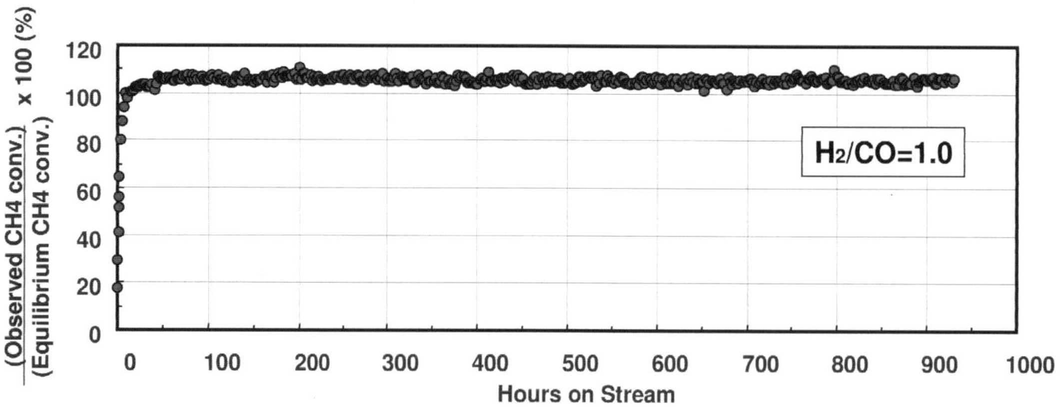

4. Application studies of new synthesis gas production process for GTL, Oxo-alcohol (n-Butanol) and acetic acid synthesis

It is considered our synthesis gas production process makes superior economic sense when the carbon dioxide is available with low cost. In the Southeast Asia and Oceania, a lot of gas fields containing a large amount of carbon dioxide are found. In these natural gas fields, our process will be successfully applied to the syngas production as feed for FT oil, DME production.

Another advantage of our process is represented in the application to GTL process integrated with LNG plants. A considerable amount of carbon dioxide is being exhausted in atmosphere form LNG plants. These carbon dioxides can be effectively utilized as a reforming agent for our process.

Our technology can also use the carbon dioxide which were exhausted from some kind of chemical plants, for example, ammonia plant and ethylene oxide as a raw material. These carbon dioxides are effectively utilized as reforming agent to produce synthesis gases for production of chemicals.

Application studies of our synthesis gas production process for GTL, Oxo-alcohol (n-butanol) and acetic acid synthesis processes were conducted in comparison with the conventional steam reforming process with CO_2 recycle. The plant capacities of GTL, n-butanol and acetic acid synthesis processes are assumed 10,000BPSD, 150,000ton/year and 500,000ton/year, respectively.

Feed gas conditions of our synthesis gas production process and conventional ones are represented on the carbon limit diagram[4)] in Fig.4. In this figure, the solid line connecting y-axis of 4.0 and x-axis of 2.0 represents the CO_2 to CH_4 molar ratio of feed gas and the straight line increasing from y-axis of 4.0 represents H_2O to CH_4 molar ratio of feed gas. The product H_2/CO molar ratios are presented dotted line in this figure. They are calculated the equilibrium gas compositions at the reformer outlet conditions of 950°C and 1.5MPa. For Fischer-Tropsch, n-butanol and acetic acid synthesis, the H_2/CO molar ratio of product synthesis gases are taken 2.0,1.0 and 0.5 respectively. It is obvious that operations on the left side of the carbon limit curve are more economical conditions because lower steam and carbon dioxide addition in feed to produce the synthesis gas for a given H_2/CO ratio. But in these conditions, product synthesis gases have high thermodynamic potential for carbon formation. In our process, feed gas conditions on the left side of carbon limit line could be selected because our catalyst has high resistance against carbon formation.

Fig.4 Feed Gas Conditions for Process Studies

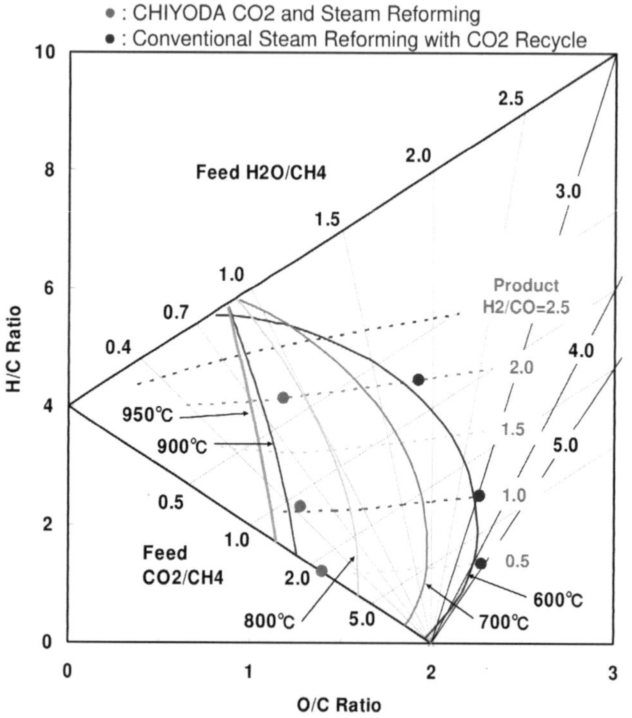

In the process studies, the configuration of synthesis gas production process is assumed as followed. The carbon dioxide in the product synthesis gas is separated by amine absorption and recycled to the feed stream. On the production of carbon monoxide for acetic acid synthesis, at first the synthesis gas with H_2/CO molar ratio of 0.5 is produced by the reformer and then hydrogen is separated by PSA from the synthesis gas.

The typical study results are represented in Fig.5, 6 and 7. In these figures, all study results of our synthesis production process represented the relative values to the conventional ones at the same H_2/CO molar ratios of synthesis gases.

Figure 5 shows the heat duty of reformer. Our reforming process represents the superior process efficiency because the fuel consumptions in the reformer are reduced in proportion to the reduction of process duty. For the case of synthesis gas production with lower H_2/CO molar ratios, the difference of heat duties between our process and conventional one becomes larger. Because larger amount of carbon dioxide and steam addition in the feed stream is required for the conventional reforming process to avoid carbon formation on the catalyst.

Fig.5 Estimated Heat Duty of Reformer

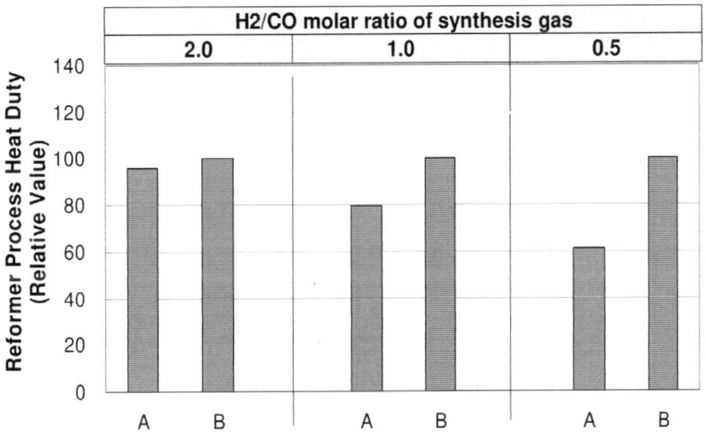

A : CHIYODA CO2 and H2O Reforming
B : Conventional Steam Reforming with CO2 Recycle

Fig.6 Estimation for amount of CO2 Emission from Synthesis Gas Production Section

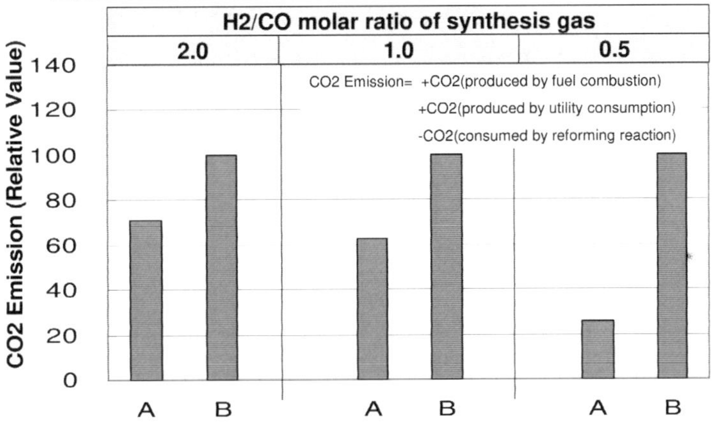

A : CHIYODA CO2 and H2O Reforming
B : Conventional Steam Reforming with CO2 Recycle

Fig.7 Plant Cost for Synthesis Gas Production Section

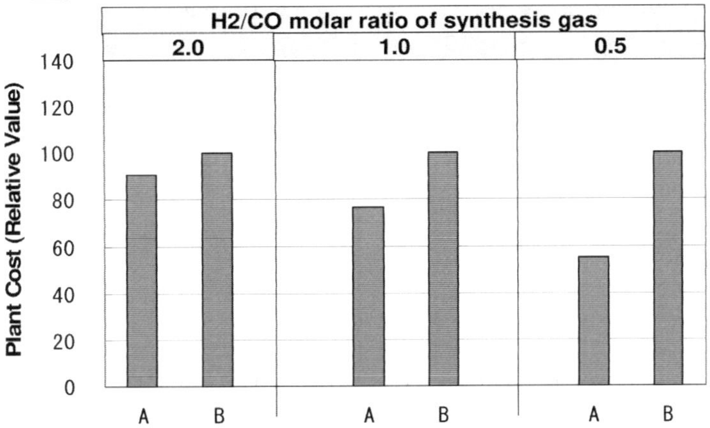

A : CHIYODA CO2 and H2O Reforming
B : Conventional Steam Reforming with CO2 Recycle

Fig.6 represents the estimated amount of carbon dioxide emission from synthesis gas production processes. The carbon dioxides produced by fuel combustion, steam generation in utility facilitiy and consumed by reforming reaction are considered to calculate carbon dioxide emissions. Our reforming process can largely reduce the amount of carbon dioxide emission by 30～70% in comparison with conventional one. Because our process is superior to conventional one in fuel and utility consumption, especially in energy consumptions for excess CO_2 removal and recycle.

Fig.7 presents the estimated plant cost for synthesis production section. Main criteria and assumptions adopted for the economic analysis are summarized as follows;

- Assumed as Grass roots plant located in Middle East.
- Capital cost includes one-time catalyst charge cost.
- Feed : Natural Gas (0.75US$/MMBTU) is assumed as representative of gas cost in Middle East.
- Feed CO_2 Gas : zero cost value
- On Stream Factor : 90% (8,000 h/year)
- Depreciation : 10% of total plant cost per year
- ROI before Taxes : 15% of total plant cost per year
- Tax and Insurance : 1% of total plant cost per year
- Maintenance cost : 3% of total plant cost per year
- Site preparation / piling / sewer / road / paving, 2-year spare parts, license fee & royalty, allowances & contingency, company overheads & profits are not included into the capital cost.

As shown in the figure, our processes represent superior economic sense to the conventional ones. This tendency became remarkable for the production of synthesis gas with lower H_2/CO molar ratio.

5. Further development of our synthesis gas production catalyst and process

Now we have been developing these catalyst and process in JOGMEC National GTL project. The GTL demonstration plant which has the synthesis oil production capacity of 500BPSD has been constructed in Japan. The demonstration test will be commenced in 2009.

6. References

1) J.R.Rostrup-Nielsen and J.-H. Bak Hansen,J.Catal.,144,38(1993)
2)Z.L.Zhang and X.E.Verykios,Catal.Today,21,589(1994)
3)M.C.J.Bradford and M.A.Vannice,Catal.Rev.Sci.Eng.,41,1(1999)
4)N.R.Udengaard, J.-H. Bak Hansen and D.C.Hanson, Oil & Gas J., Mar.9(1992)

25a Evolution and Revolution: Process Innovations In Aromatics Technology

Gavin P. Towler, UOP, 25 E Algonquin Rd, Des Plaines, IL 60017-5017

When AIChE was founded, there was already a substantial market for aromatic compounds such as benzene and phenol, which were primarily recovered as byproducts of coal gasification. The demand for aromatics and their derivatives increased enormously in the second half of the twentieth century as new polymers were commercialized. The invention of catalytic cracking and later catalytic naphtha reforming made available cheaper benzene, toluene and xylenes (BTX) from petroleum feedstocks and laid the foundation for petroleum refiners to enter the aromatics business.

The production of aromatics from petroleum feedstocks has followed a pattern of "punctuated equilibrium", with periods of steady increase in yields and plant sizes followed by radical breakthroughs that dramatically improved process economics. The successful commercialization of separation processes such as Sulfolane and Parex allowed desired compounds such as benzene and paraxylene to be recovered at higher purity and higher recovery. At the same time, new zeolite catalysts enabled the isomerization of xylenes and toluene disproportionation and transalkylation, allowing aromatics producers to tailor the yields of products to better match the market demand.

Aromatics production continues to evolve. In the past decade, new processes have been commercialized to allow aromatics production from cheaper feedstocks such as light hydrocarbons (Cyclar) and heavier feedstocks such as light cycle oil (LCO-X). The construction of bigger, more heat-integrated complexes is also driving continued innovation as we move into AIChE's second century.

25b A Process History of Lexan Polycarbonate Resin

William Hollar[1], Robert R. Gallucci[1], James Hoover[1], Donald Paciorkowski[1], and Ignacio Vic Fernandez[2]. (1) Sabic Innovative Plastics, One Lexan Lane, Mount Vernon, IN 47620, (2) SABIC Innovative Plastics CR Cartagena, Alhama de Murcia Km 13, La Aljorra, Cartagena, 30930, Spain

Lexan aromatic polycarbonate (PC) based on bisphenol A (BPA) was originally invented in the 1950's. Its unique physical properties include transparency, extreme toughness and elevated temperature property retention. These attractive properties have resulted in market growth to greater than 3 million metric tons/year globally. Polycarbonate applications are widespread and include automotive, media, glazing, lighting and electronic equipment along with a myriad of consumer applications. Chances are excellent that you own several items made of polycarbonate.

This presentation will focus on the history and evolution of the Lexan polycarbonate manufacturing process and its tie to product performance and applications. Originally based on interfacial phosgenation processes, polycarbonate is also produced by melt transesterification processes. The key process features and a comparison of the two processes will be discussed.

Finally, as the market for polycarbonate matures, SABIC Innovative Plastics has increased its focus on producing novel Lexan polycarbonate copolymers with enhanced properties for demanding product applications. This has resulted in new process challenges and breakthroughs for commercial production. An example of a Lexan polycarbonate co-polymer process will be reviewed.

LEXAN is a registered trademark of SABIC Innovative Plastics IP B.V.

262a Hydrotreating Reaction Performance In Small Catalyzed Channels

Wei Liu, Energy and Environmental Directorate, Pacific Northwest National Lab, 902 Battelle Boulevard, MSIN:K2-12, Richland, WA 99352

This work aims to address fundamental feasibility of conducting hydrotreating reactions in structured catalyst channels in general and in ceramic monolithic catalyst modules in particular. CoMo/gamma-alumina hydrotreating catalysts made in the monolith structure form of channel size of about 1mm are tested with real-world refinery feedstock under steady-state reaction conditions within the range of typical hydrotreating process operations. The reaction performance of the same catalyst material in the monolith-structured form is compared to that in the trickle bed configuration of crushed particles and cylindrical catalysts. The exceptionally high HDS and HDN activity is observed in the structured catalyst channel. The reaction activity of the structured catalyst is a few times that of the trickle bed activity. Such a large degree of the activity enhancement obtained by changing the catalyst bed structure (not intrinsic catalyst material) is explained by more effective catalyst utilization in the confined reaction channels than in the random-packed trickle bed.

262b Process Simulation of a Pilot Case: Production of Perfume-Containing Microcapsules

Laureano Jiménez-Esteller, Department of Chemical Engineering, University Rovira i Virgili, Av. Països Catalans, 26, Tarragona, 43002, Spain and Rodolfo V. Tona Vásquez, Chemical Engineering, University Rovira i Virgili, Av. Països Catalans, 26, Tarragona, 43002, Spain.

1. Introduction On the basis of a radical change in the paradigm microtechnology may help to build the foundations for a new generation of chemical production plants for high added value products. Microtechnology is able to achieve process and products improvements by orders of magnitude, rather than percentage points on key sectors (pharmaceuticals, additives, pigments, coatings, food additives…). The idea behind microdevices is the combination of two concepts: · Intensification: produce more with less energy, less solvent, less inventory and reduced transport, thus reducing the environmental impact. · Miniaturization: supply better products with smaller volumes, higher efficiency and greater precision. The need for microsystem modeling arises for two reasons: (i) to allow the rapid penetration into the market it is essential that the design, manufacture and test loop is kept to a minimum; (ii) device models can lead to a higher understanding of microdevice operation. To this end, model complexity (many of the new microsystems including several sensor types along with their relevant conditioning and read out electronics) and material properties (surface effects and microscopic thermal properties) are among the most important. **2. Perfume-Containing Microcapsules (PmC)** The industrial case is the synthesis and characterization of perfume-containing microcapsules (PmC). PmC are of great interest to the industrial partner as they offer a mechanism for the efficient deposition of perfumes as well as provide a long-lasting fragrance benefits. Perfume deposition on textiles is an inefficient process (» 90 % of perfume added to a detergent is lost during washing). In addition, perfumes have to be very fabric specific and this limits the possible fragrances. Encapsulation of perfume into a microcapsule helps deposition because, with careful control of the size, PmC become entrapped into the cloth fibers during washing and resist being flushed away. The production of PmC using microtechnology is at the laboratory stage. The continuous encapsulation technique is phase inversion precipitation The process occurs in a two steps procedure (Figure 1). The polymeric solution and the continuous phase is fed to the first microreactor. The continuous phase is immiscible with the solvent and is not a non-solvent for the polymer, and therefore the product stream is an emulsion containing polymer and perfume microdroplets. In the second microreactor the non-solvent phase is added, and precipitation of the polymer occurs. As a result, microcapsules are produced with the perfume encapsulated. To avoid any operational problem, and according to experimental data, the purity specifications of raw materials is very high (99-99.9 % w/w). In addition, the polymeric phase stream must be free of water [1]. **3. Integration of Process Modeling Approaches** The MICAP® module is used to perform the simulation of the PmC. The MICAP® tool, developed in Matlab® [2, 3], is used to link the modeling approaches: 1. Molecular simulation to study the micellization process, the precipitation of the polymer, size of porous and the morphology of the microcapsules. 2. Computer fluid dynamics to predict the flow dynamics, the mixing, and mass/heat transfer processes in the microreactor. 3. Process simulation to predict the continuous production of the PmC, including the solvent recovery system. 4. Life cycle assessment to deal with the environmental impact and consider the sustainability. This work is focused at the process simulation level. Different modeling tools were used, as commercial process simulators are not prepared to efficiently handle microtechnology problems. The tasks are divided in two levels: (a) encapsulation of the MICAP® within Aspen Plus®; (b) PmC separation and recovery of the solvents. **3.1. Encapsulation of MICAP® and Aspen Plus®** Current process modeling and simulating tools offer a broad variety of functionalities to develop reliable models. The process simulator selected is Aspen Plus®. The tool does not offer functionalities for handling microstructured unit operations, and therefore the modeling requires an additional effort to overcame some difficulties [4]. Aspen Plus® and Matlab® were linked using Active X and COM technology [5, 6], thus integrating both tools in the calculation procedure. The problem of this integration is that both calculations tools are involved in several nested loops (Figure 3) giving rise to convergence problems. As Aspen Plus® is a sequential-modular simulator, the modification of the process topology does not affect the overall convergence performance of the system. In this way, the architecture and transfer procedure of the modules is build to promote scalability, re-usability and flexibility of the model A Basic Unit Operation (BUO) was created within Aspen Plus®. The data from MICAP® is send to the BUO and data calculated from Aspen Plus® may be recovered through this BUO. The BUO acts as an encapsulate version of MICAP® within Aspen Plus® and, consequently, as a microdevice unit within the simulator (Figure 2). **4. Perfume-Containing Microcapsules Separation and Solvent Recovery** The microdevice output stream contains the PmC with cyclohexane, DMF, water and perfume (vanilla). Although this separation has not been experimentally tested, it is based in the information available. Figure 3 shows a complete diagram of the process. **4.1. Recovery of cyclohexane** Cyclohexane is highly immiscible with water and DMF [7, 8], and the modeling of the decanter is straightforward. Then, the adoption of microstructured technology does not offer any a priori advantage. Therefore, the model proposed a scheme with microdevices working in parallel (N = 238) and the resulting stream is processed in a decanter, where the organic phase (cyclohexane), and the aqueous phase (PmC, DMF, water and perfume) are obtained. The number of microdevies can be modified by the user, and this change of scale allows to combining multiscale process units with microstructured elements. **4.2. Recovery of microcapsules** The average size of microcapsules in the laboratory is over 6 microns with a very small percentage under 4.5 microns (2% in weights). Then, a screen filter to retain particles (efficiency » 95-99 % at 4.5 microns) is enough for PmC recovery. To mimic the production line PmC must be preserved in a gel media (» 20 % liquid). In addition, the porous structure of the microcapsules retain some liquid (» 5 % w/w). **4.3. Recovery of water and DMF** The filtrate liquid contains DMF, water and a small fraction of perfume. The separation of water and DMF mixtures is the testing system for pervaporation units [9, 10]. Due to miscibility, perfume is recovered with the DMF (free-water). **4.4. Minimizing the environmental impacts** A key aspect in the pilot plant tests is the reuse of solvents, leading to a minimization of wastes and raw materials. Nevertheless, a life cycle assessment (LCA) is performed following the cradle to grave approach using the Ecoinvent [11] database and SimaPro® [12]. The aggregation of the impact categories into a single value (Eco-indicator 99) improves its value for the decision makers. Most of the environmental load is assigned to the DMF production, and the residual streams containing DMF. Effort is needed to minimize this stream or select an alternative solvent. **5. Conclusions** It is clear that at present there are many industrial applications of microdevices (biosensors, microprocessors…), the state of the art in chemical engineering is far from the one achieved in other areas. The end goal of the IMPULSE project ("fab-on-a-chip") is far from achieved, as microdevice has a clear advantage in some market niche (e.g. microreactors), while separation technology is far from developed. In addition to model microsystems, experimental data based on real behavior is required to tune the model predictions (drag-out, efficiencies…).

6. Acknowledgements This work has been funded by the IMPULSE EU project (NMP2-CT-2005-011816).

Literature [1] Production of capsules using modified lab reactor - Deliverable D3.2f. IMPULSE. ETSEQ, February 2007. [2] C. Torras, and R. Garcia. The [MICAP] simulator. Registered software T-177/07. [3] Torras_MICAP_Flyer.pdf and Torras_MICAP_Oral.pdf (available at the IMPULSE website). [4] Aspen Plus® 12.1. User Guide. Aspen Technology Inc., Cambridge (MA), USA. 2003. [5] A. Bojarski, L. Jiménez, A. Espuña and L. Puigjaner. European Congress of Chemical Engineering. Copenhagen, 2007. [6] Matlab® 7.0. External Interfaces. The MathWorks, Inc. Natick (MA), 2007. pp. (8-3) - (8-129). 2007. [7] M.L. Campbell. "Ciclohexane" in Ullmann's Encylopedia of Industrial Chemistry (6th Ed). John Wiley & Sons, Inc., 2007. [8] H. Bip and H. Kieczka. "Formamides" in Ullmann's Encylopedia of Industrial Chemistry (6th Ed). John Wiley & Sons, Inc., 2007. [9] H.K. Lee, J.Y. Kim, Y.D. Kim, J.Y. Shin and S.C. Kim. Polymer, 42 (2001) 3893. [10] D.Anjali Devi, B. Smitha, S. Sridhar and T.M. Aminabhavi. Separation and Purification Technology, 51 (2006) 104. [11] Swiss Centre for Life Cycle Inventories, The Ecoinvent database, (2006). [12] Pré-Product Ecology Consultants, Introduction to LCA with SimaPro® 7, Pré-Product Ecology Consultants, (2006).

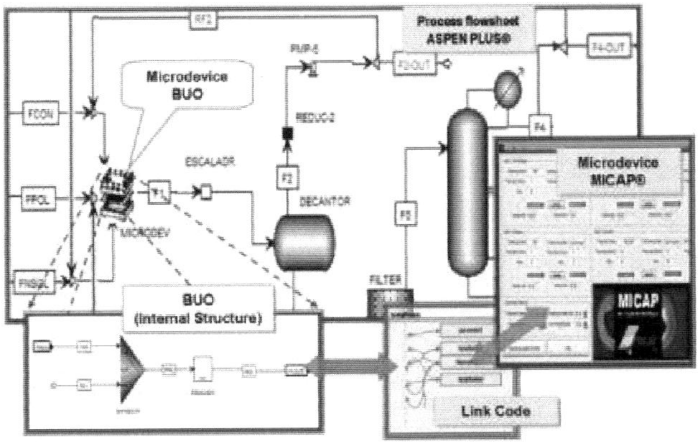

262c Gas-Liquid Mass Transfer in a Rotating Horizontal Cylinder

Gangadhara Raja Muthinti[1], Roshan J. J. Jachuck[2], and Srinivasan Ambatipati[2]. (1) Chemical Engineering, Clarkson University, CAMP 231, 8 clarkson avenue, Department of chemical and biomolecular engineering, Potsdam, NY 13699, (2) Chemical and Biomolecular Engineering, Process Intensification and Clean Technology (PICT) Group,Clarkson University, PO Box 5707, Potsdam, NY 13699

Rotating horizontal cylinder is one of the upcoming intensified modules in the field of process intensification. There is not much literature available on the mass transfer studies in a rotating tube. Thin hydraulic path lengths are obtained in a rotating tube in comparison to the conventional mass transfer equipments. This enhances the rate of mass transfer by orders of magnitude which is one of the important aspects of process intensification. The main aim of this paper is to determine the mass-transfer characteristics of the horizontal rotating tube in continuous mode and compare it with other intensified modules. Experiments have been carried out with deionized water as liquid phase and pure nitrogen as the gas phase to study the stripping of dissolved oxygen from liquid to gas phase.

HIGH PRESSURE MICROFLUIDICS FOR NANOMATERIALS SYNTHESIS

Samuel Marre[1], *Jongnam Park*[2], *Moungi G. Bawendi*[2], *Klavs F. Jensen*[1]
[1]*Chemical Engineering Department and* [2]*Chemistry Department, Massachusetts Institute of Technology, Cambridge, MA, USA*

Continuous synthesis of nanomaterials in microfluidic devices offers several advantages over conventional macroscale chemical processes including enhancement of mass and heat transfer, reproducibility, potential for sensor integration for in situ reaction monitoring, rapid screening of parameters, and low reagent consumption during optimization. The synthesis of inorganic nanostructures generally requires high temperatures to decompose metal organic precursors and to generate nucleation followed by further growth. However, limitations on the number of compatible chemistries and the limited number of available high boiling point solvents have been a major obstacle in the rapid adoption of microreactors as a universal platform for nanomaterials synthesis. One way of improving syntheses in microreactors is to perform experiments at high pressure. Indeed, at sufficiently high pressure, virtually any common solvent, precursor, and ligands will remain either liquid or become supercritical (sc) at the temperatures required for the syntheses.

In this context, we have developed high pressure high temperature continuous-flow Silicon-Pyrex microreactors for the synthesis of nanomaterials. The microreactors consisted in an injection zone maintained at room temperature and in a reaction zone under temperature. The two zones are separated by a thermally isolating halo etch that allowed for a temperature gradient of 250°C/cm. High pressure modular compression fluidic connections were realized by compressing the microreactor between two stainless steel parts using viton O-rings. In this configuration, the set-up allows reaching high pressure (up to 25 MPa) and temperature (up to 500 °C in the heated section) (Figure 1).

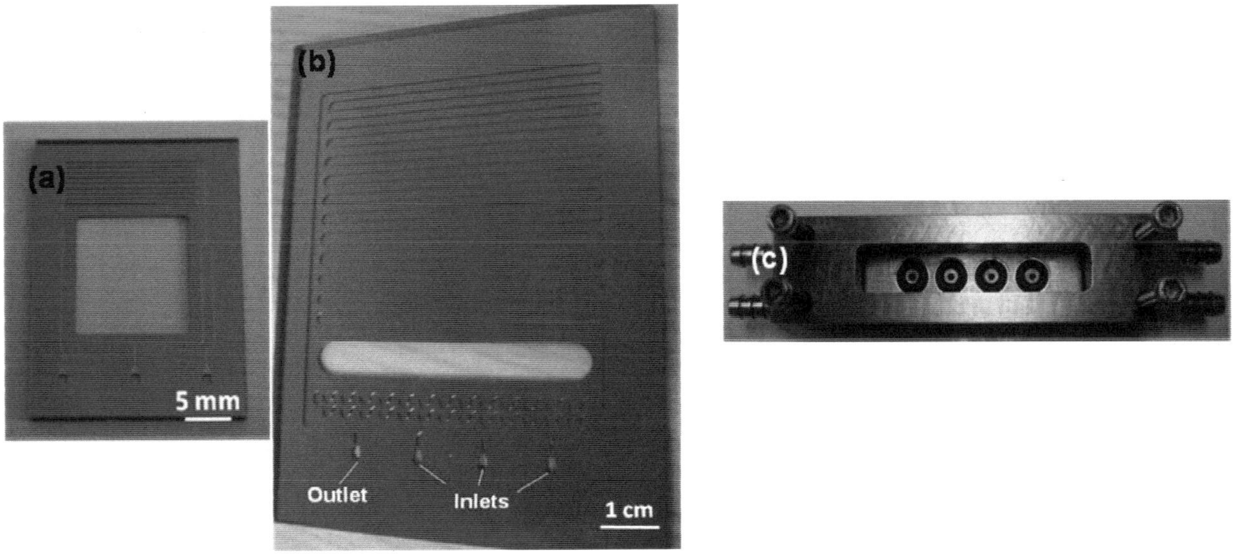

Figure 1. High pressure microreactors for nanomaterials synthesis: (a) 4.5 μL microreactor, (b) 100 μL microreactor and (c) modular fluidic connections.

Model nanomaterials (CdSe QDs, Pd nanoparticles, iron oxide nanocrystals) were synthesized to demonstrate the process under pressure, while tuning several parameters: solvent, temperature, residence time or concentration. In the case of QDs, we demonstrate that the use of supercritical fluids, in particular, shows strong influence on both the concentration of the QDs and on their size distribution, leading to small Full Width at Half Maximum (FWHM) of the emission peak.

CdSe quantum dots were obtained either from either squalane, dodecane, toluene or supercritical hexane, for T ranging from 210°C to 310°C at 50 bars. Iron oxide nanocrystals were synthesized in supercritical toluene at 325°C and 100 bars. Palladium supported nanoparticles on silica were synthesized in supercritical CO_2 – ethanol media at 200°C and 130 bars.

The use of high pressure high temperature microreactors demonstrated in this work opens new routes for nanomaterials synthesis in microfluidic devices by enlarging the set of solvents and phases (liquid, gas, supercritical) available.

262e Single-Phase Rimming Flow In a Rotating Horizontal Cylinder

Saravanan Suppiah Singaram, Chemical and Bio-molecular Engineering, Clarkson University, 8 Clarkson Avenue, PO Box 5707, Potsdam, NY 13699 and Roshan J.J Jachuck, Process Intensification & Clean Technology (PICT) Group, Department of Chemical Engineering, Clarkson University, PO Box 5705, Potsdam, NY 13699.

Process intensification is based on the idea that the transport rates are dramatically influenced when chemical processes are carried out in smaller path lengths and also in a continuous mode of operation. In this work, a special type of coating flow that leads to a thin film of liquid onto the inner surface of a horizontally rotating cylinder is studied. This type of flow is typically called rimming flow. At low angular velocities, the flow settles into a steady flow with a front where the coating film coalesces with the pool at the bottom of the cylinder. At higher angular velocities, a nearly uniform thin film (annular flow) is formed inside the rotating cylinder. There is extensive experimental as well as theoretical data available in literature on rimming flow of viscous liquids in partially-filled (no outflow) rotating cylinders. The aim of this work is to study rimming flow in a rotating cylinder that has a continuous outflow of liquid. Film thickness measurements (in the annular flow regime) are important to predict the transport performance of the developed flow system. Rimming flow of water is studied in this work and the film thickness measurements are done using an interferometric technique. The evaluation of the measurement technique and the experimental film thickness measurements will be presented. A correlation using dimensional analysis and also some preliminary theoretical work will be presented.

262f Dilue Acid Hydrolysis of Cellulosic Wastes Using Field Effects

Hariprasad Amanapu and Prof. Roshan JJ Jachuck. Chemical Engineering, Clarkson university, 100 Market street,Appt # 16, Potsdam, NY 13676

The use of microwaves is making inroads into many processes such as chemical synthesis and firing ceramics. Previous research on microwaves has revealed the significant effect of microwaves on reaction acceleration; yield improvement and selectivity enhancement through different reaction pathways. Fundamentally, microwaves heat things differently than conventional means. The present study is an attempt to apply microwaves and study the dilute acid hydrolysis of cellulosic wastes. Product is analyzed with HPLC with refractive index detector. The analysis of results reveals that dilute acid hydrolysis of cellulose follows first order kinetics. The rate constant and activation energies have also been estimated.

373a Kamlesh Bhatia: A Career In Innovation

Surita R. Bhatia, Chemical Engineering, University of Massachusetts, Amherst, 686 North Pleasant Street, Amherst, MA 01003-9303 and Sujata K. Bhatia, Materials Science and Engineering, DuPont Central Research & Development, Experimental Station, E328, P.O. Box 80328, Wilmington, DE 19880-0328.

We give an introduction and overview of the career of Dr. Kamlesh K. Bhatia, highlighting his major accomplishments and his influences on current and future generations of chemical engineers.

373b Yoga Meditation and Black Belt Action In Process Research and Development

George Liebermann, Xerox, 2660 Speakman Dr., Mississauga, ON L5K 2L1, Canada

This paper, in honor of Dr. Kamlesh K. Bhatia, will discuss the opportunities offered by a judicious use of creative thinking as well as of structured, delivery oriented protocols in process research and development. Generation of ideas is usually expected to be the result of an individual mind, however it is also many times the aftermath of interactions of several people. The creation of ideas does not always guarantee a successful outcome. The selection of the best idea to invest in and work on can provide challenges and heuristic know-how, intuition and some "tools" can be used in many circumstances. The development of the idea (invention) to a practical new process (innovation) requires a disciplined approach, however creativity will still play a major role.

The above points will be considered in detail in a number of examples from specialty organic chemicals and polymer processes.

373c A Novel Method for Enhanced Heat Removal from Closed Systems

Ranga Narayanan and Jonathan M. Volk. Chemical Engineering Department, University of Florida, Gainesville, FL 32611

Removal of heat from confined cabins and space habitats is of immense importance to NASA in order to ensure proper thermal control. The design of important thermal subsystems for future applications as in boilers, condensers, evaporators, heat exchangers etc will involve complicated fluid flow systems. Enhanced heat removal is needed when high heat generation occurs as in materials processing experiments. Removal of this heat is important for proper thermal control for human habitation and instruments in an extra terrestrial space environment. Such technology also has spin-off uses in Earth applications as in the thermal stability of chemical reactors, removal of heat from electrical devices, and other enclosed environments such as submarines and underground facilities.In our talk we will discuss a novel method for the enhanced heat removal from an enclosed system. This method is independent of gravity and can work in single or two phase regimes. This novel idea incorporates the use of pulsatile flows to enhance the heat transport. The physical idea behind this is that if a heat source is connected to a heat sink via a fluid and if the fluid were oscillated, the convective motion will bring about sharp spikes in the velocity profile which in turn will transport the heat faster in both the radial and axial directions. It has been shown that through experiment and the theory that these pulsatile flows increase the heat transport of a system by as much as three orders of magnitude compared to ordinary conduction. Fluids as heat transport carriers are far better than metals because of the reduced added mass in space flights, particularly during lift-off. Our talk will focus on the physics of the enhanced heat transport and the verification of the theory by defined experiments.

373d Design and Process Control Issues In Nanomanufacturing

Babatunde A. Ogunnaike, Chemical Engineering, University of Delaware, 150 Academy St., Newark, DE 19716

This presentation will raise awareness of some key process issues involved in answering the following primary question: in manufacturing products designed for specific end-use applications (e.g. polymer nano-composites); or in the production of self-assembled nanomaterials with desirable structural characteristics (particle size, shape, density, and spatial self-organization); what strategy is required for achieving effective design and control of product properties and for assuring acceptable end-use product performance?

Development of Current and Next-Generation Biomedical Materials

Sujata K. Bhatia, Materials Science and Engineering, DuPont Central Research & Development, Experimental Station, E328, P.O. Box 80328, Wilmington, DE 19880-0328

Biomaterials science is a multidisciplinary endeavor incorporating chemical engineering, medicine, biology, chemistry, materials science, bioengineering, and biomechanics. The past few years have witnessed an explosion in the field of biomedical materials, with an expansion of both the compositions and the applications of medical implant materials. As the prevalence of chronic diseases such as diabetes, cardiovascular disease, and neurodegenerative disease increases, there will be an even greater need for innovative biomaterials. This talk reviews the current status of the field of biomaterials, and highlights new developments in biomedical materials. The talk will provide an overview of medical applications of biomaterials, and will describe current classes of biomaterials, including metals, ceramics and glasses, and polymeric materials. The talk will then discuss the next generation of biomedical materials, including surface-modified biomaterials, smart biomaterials, bioactive materials, biomimetic materials, patterned biomaterials, and tissue engineering and regenerative medicine. There are numerous opportunities for chemical engineers to develop novel biomedical materials that will improve and save lives.

Reference:

S. K. Bhatia and S. R. Bhatia, "Biomaterials," *Encyclopedia of Chemical Processing, 2nd Ed.* (2005).

373f Process Development - the Role of Engineering Evaluations

Bruce M. Vrana and Robert A Cantini. DuPont, 1007 Market St, D5100, Wilmington, DE 19898

Process development is a crucial activity for organizations. The role of Engineering Evaluations, as practiced at DuPont, in guiding process development, will be discussed. There may even be an anecdote from Kamlesh Bhatia's career.

373g The Yoga of Process Development

Kamlesh Bhatia, American Institute of Chemical Engineers, 18 Sheldrake Road, Newark, DE 19713

Process Development is analogous to Yoga. It is a 'Yoga' – a disciplined process for integrating the essential components to create a vision of the 'Whole' and experiencing the 'Joy' of professional satisfaction. In this talk, I will point to some ways for practicing this 'Yoga' with ease and illustrate with examples from personal encounters.

438a Conceptual Design of Biobutanol Production Processes

Arjan Oudshoorn, Adrie J.J. Straathof, and Luuk A.M. van der Wielen. Department of Biotechnology, Delft University of Technology, Julianalaan 67, Delft, 2628BC, Netherlands

Increasing interest in sustainable industry, renewable feedstock utilization and increase in oil prices has led to renewed attention for Acetone Butanol and Ethanol (ABE) fermentation from industry and academia, in particular for 1-butanol.

Our aim is to develop an efficient butanol production process taking into account Clostridial fermentation as well as downstream processing.

The biological production of butanol by ABE fermentation has been shown to provide a total solvent concentration in the fermentation broth of around 20 g/L. The product concentration is limited mostly by product inhibition of butanol. Product removal during fermentation will increase the productivity and help the overall economy of the production process. An efficient product removal operation will be able to minimize the production costs, including the fermenter costs, by maintaining a non-inhibiting butanol concentration in the fermenter. Therefore the key downstream step in the process is the separation of cell-free fermentation broth into a small butanol-rich and a large butanol-poor fraction.

A large array of options for butanol recovery from aqueous solutions exists. It is difficult to compare these without having to design all the potential integrated process options on the same basis in full detail. Therefore a simpler method to rank the possible recovery options has been developed. This method focuses on the selectivity between butanol and water of the recovery, the resulting mass flows, and the corresponding energy consumption. Energy consumption is a key operational cost factor with respect to the recovery.

According to this analysis, one of the best options for recovery is adsorption of butanol by high silica zeolites. Potentially this can directly lead to relatively pure butanol streams upon desorption. Scaling-up an adsorption/desorption operation leads to a number of complications that will be illustrated.

The design of an overall process including fermentation and adsorption shows that the process can be economically feasible, depending on the prices taken for the raw materials and the product. Important design issues that will be discussed are:

• Continuous fermentation

• Cell recycle

• Minimizing acetone and ethanol by metabolic engineering

• The role of H2 in the process

This study is financially supported by the Netherlands Ministry of Economic Affairs and the B-Basic partner organizations (www.b-basic.nl) through B-Basic, a public private NWO-ACTS programme (ACTS: Advanced Chemical Technologies for Sustainability).

438b Flare Minimization for Chemical Process Industry Sustainability

Qiang Xu, Kuyen Li, Chao-Wei Liu, and Xiongtao Yang. Department of Chemical Engineering, Lamar University, Beaumont, TX 77710

During chemical plant turnaround operations (plant startup and shutdown) and process upset management, off-spec product streams will be produced, which usually have to be sent for flaring. Flaring is a passive and end-of-pipe means to protect personnel and equipment safety in chemical process industry (CPI). However, flaring inevitably generates huge amount of CO, CO2, NOx, and some extremely hazardous volatile organic compounds (VOCs), which will cause severe air pollution problems and negative societal impacts. Meanwhile, it also results in tremendous raw material and energy loss that could generate much needed products from the industry. Thus, flare minimization has great significance on environmental, societal, and economic advancement toward the sustainability of chemical process industry.

Facing this challenge, one economic and effective way to minimize flaring during the plant turnaround operation is to use plant-wide dynamic simulation to critically check operating procedures before the real operation. The turnaround procedures are the steps taken by the plant operating personnel to operate a plant. Each of these steps involves the manipulation of the process plant through sophisticated control systems. The plant-wide dynamic simulation can identify benchmark allocations of flaring emissions during plant start-up, shutdown, and normal upset maintenance. It targets the emission source reduction during the operation; instead of expensive process modifications for industrial flare system.

The methodology starts with a development of steady-state simulation model for the target process area. The steady state simulation model is validated by plant data, which is from DCS historian. This validated steady state model is then transferred to a dynamic model by following the updated plant P&ID to set up the control strategy and input the control parameters and equipment dimensions. Again, the plant historian data are used to validate the plant-wide dynamic simulation model. After validation, the plant-wide dynamic model will be gradually turned to the ready to turnaround operation status. With the startup procedures, dynamic simulation will be conducted to identify possible unstable or risky operation conditions. When such unstable or risky conditions are identified, the plant-wide dynamic simulation model can be used to find new stable operation conditions. With the cooperation of the plant operation group, optimal plant turnaround operations can also be identified. The plant-wide simulation is based on first principles. Therefore, the developed methodology is applicable in general and not limited to one particular plant or process.

The dynamic simulation gives an insight into the process dynamic behavior that is not apparent through the use of steady-state simulation or industrial expertise alone. This information is crucial for plant turnaround operations in order to minimize the flaring. The efficacy of this methodology and its economic and environmental benefits for industrial sustainability are demonstrated by real field startup tests.

438c Process and Economic Simulation of Wastewater Plants for Lipid Feedstock Production through Use of Superpro Designer

Stephen Dufreche[1], William French[2], and Rafael Hernandez[2]. (1) Dave C. Swalm School of Chemical Engineering, Mississippi State University, 330 Swalm President's Circle Dr., Mississippi State, MS 39762, (2) Chemical Engineering, Mississippi State University, 330 Swalm President's Circle Dr., Mississippi State, MS 39762

The biofuels market is currently expanding at a remarkable rate, but a major concern is the continued availability of feedstock needed to meet demand. Previous research by the Renewable Fuels and Chemicals Laboratory (RFCL) at Mississippi State University has shown that microorganisms grown at municipal wastewater treatment plants are a viable source of lipid feedstock for biofuel production. While traditional wastewater plants do produce significant amounts of lipid, the standard plant configuration is designed for minimal production of biomass due to environmental reasons. New or modified plant configurations could give increased lipid yield through changes in process parameters or introduction of new organisms.

The SuperPro Designer simulation package by Intelligen, Inc. was used to simulate wastewater plant configurations for lipid production. Parameters examined include indigenous microorganisms in the wastewater versus the addition of oleaginous yeast for lipid production, pretreatment of the sludge with ozone, addition of acid hydrolysate sugars as a carbon source, and different wastewater feed sources.

Data presented from the simulation models include retention times and sizing data for equipment, characteristics of the effluent streams for EPA compliance, and overall lipid yield. Data on the economics of each design will also be presented.

438d Solar-Thermal Processing for Sustainable Production of Chemicals and Fuels

Christopher Perkins[1], Paul Lichty[2], and **Alan W. Weimer**[2]. (1) Department of Chemical and Biological Engineering, University of Colorado at Boulder, 1111 Engineering Drive, Boulder, CO 80309-0424, (2) Department of Chemical and Biological Engineering, University of Colorado, 1111 Engineering Drive, Boulder, CO 80309-0424

Concentrated sunlight can be used to directly drive high temperature chemical reactions such as metal oxide reduction for water splitting thermochemical cycles and biomass conversion to syngas. Annual average solar to reactor efficiencies approach 45% for reactor operating temperatures in the range of 1300 to 1800 C. Materials challenges dominate the process for temperatures above 1300 C, while operation at or below 1300 C can be carried out using high temperature metal alloys.

The specific example of hydrogen production using concentrated sunlight will be presented for both water splitting thermochemical cycles and biomass conversion followed by water gas shift. Experimental results will be presented for both cases along with an economic profitability analysis. Design considerations will be discussed for solar fields and receivers. Critical challenges facing solar-thermal processing will be discussed.

Photocatalytic Concrete Slab for Passive NOx Pollution Control

M.Palla, A. Vaddi, R. Asapu, D.Chen, K. Li, and R.Yuan, Photocatalysis & Green Composite Labs, Lamar University, Beaumont, Texas 77710

Extended Abstract

Introduction

NO_x emissions have become a constraint for sustainable economic development due to ozone non-attainment [1-5]. Under light illumination over a photocatalyst free radicals are generated to oxidize or decompose pollutants. While volatile organic compounds (VOCs) are oxidized to water, CO_2, and chloride, NOx are oxidized to nitric acid, neutralized by the alkaline materials in concrete, and washed away by rain [6-9]. Ozone can be decomposed to oxygen over photocatalysts. Recent developments in Europe Japan, Hong Kong, and US have demonstrated the interest in deploying photocatalytic technologies for environmental remediation. The photocatalytic coating has potential to economically reduce nitrogen oxides (NOx) emissions at ambient conditions nearly maintenance-free [10-12].

We report here the application of photocatalytic coatings on a concrete slab to simulate road pavements for passive NOx pollution control. Catalysts tested include Degussa P-25 TiO_2 and TiO_2 - boosted with $BaTiO_3$, Nd_2O_3, and Fe_2O_3 [13-16]. Both NO and NO_2 oxidation including the effects of inlet NO concentration, residence time, and humidity are investigated. The removal efficiencies are in the range of 86-93% from 200 ppb-2000 ppb NO. Relative Humidity (RH) is seen to have some beneficial effect at <35% RH while lowering NO conversion by ~10% at > 35% RH. The additives Nd_2O_3, Fe_2O_3, and $BaTiO_3$ are observed to depress the NO_2 generation while improving NO conversion when using fiber-optic as a support. When tested on catalytic concrete slabs, however, Degussa P-25 offers the best NO removal, even though Fe_2O_3 and Nd_2O_3 offer better NO_2 suppression. The optimal TiO_2 to cement ratio is of interest because TiO_2:SiO_2 photocatalyst shows twin peaks in their activity in atrazine oxidation and the fact that cement also contains Fe_2O_3. A CFD simulation of NOx plumes from a tail pipe shows the likely distribution of concentration and residence time due to dispersion; that, in turn, can provide a useful guide for future lab investigations.

Experimental Setup - Catalyst-Coated Concrete Photoreactor (CCP)

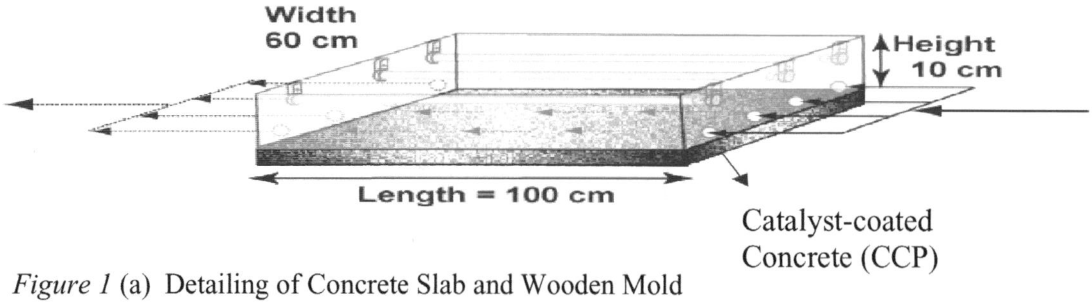

Figure 1 (a) Detailing of Concrete Slab and Wooden Mold

The catalyst-coated Concrete Photoreactor (CCP) is shown in Figure 1 (a)-(c). Three 50W (800 mA) fluorescent lamps (Philips, Model # F36T12/D/HO) are used with a total radiation of 250 W/m^2 as compared to a high-noon solar radiation of 1.44 Langleys/min (1004.18 W/m^2). The Teledyne API NOx ambient level/trace level analyzer (Model # 200EU 20B-41) is used for NOx sampling.

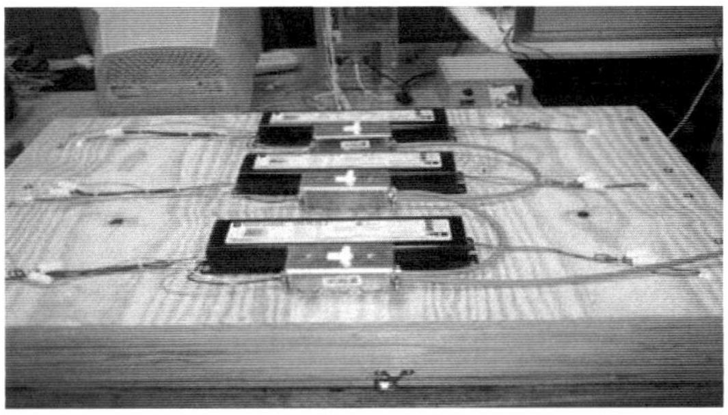

Figure 1 (b) Lamp Fixture Assembly

Figure 1(c) Simulated Solar Radiation

The operating conditions of the NO oxidation over a catalytic cement surface are given below: Residence Time: 20 min; Light Source: 50W Fluorescent Light ; Reactant: NO (inlet concentration 2.5 ppm); Catalyst on Concrete Surface: TiO_2 Degussa P-25; Catalyst Loading: 33.33 mg/cm^2.

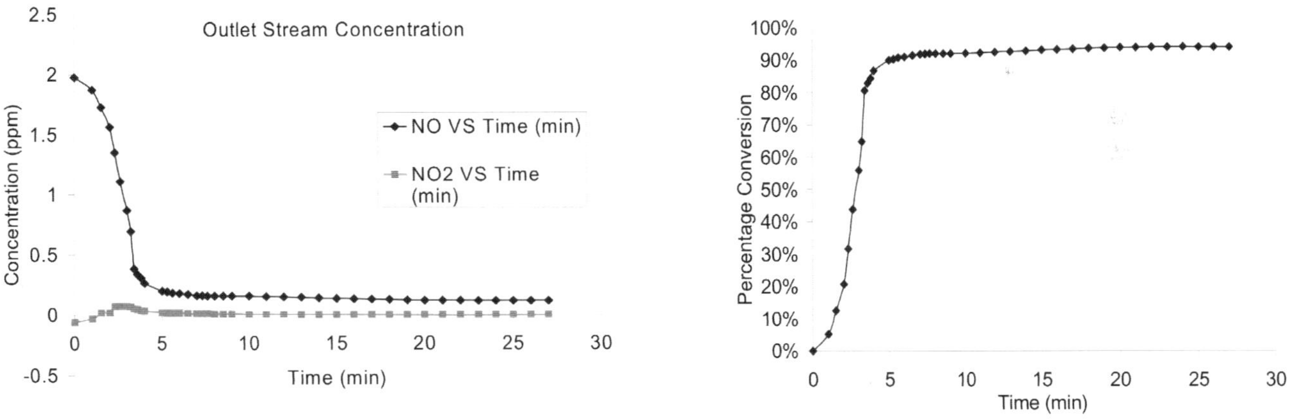

Fig. 2 (a) NO/NO2 concentrations versus time. (b) NO conversion versus time

It can be seen that the NO conversion reached 90% within 5 min. and climbed to stay above that level throughout the experiment. The NO_2 level is almost non-detectable except during a transient period within the first 5 minutes, Fig. 2(a)(b). The NO conversion increases from 86-93% with residence time (RT) from 5 - 20 min. Inlet NO concentration has little effect to the steady state NO conversion from 200-2000 ppb. Therefore, the NO conversion is insensitive to the inlet concentration.

Relative Humidity (RH) Effect

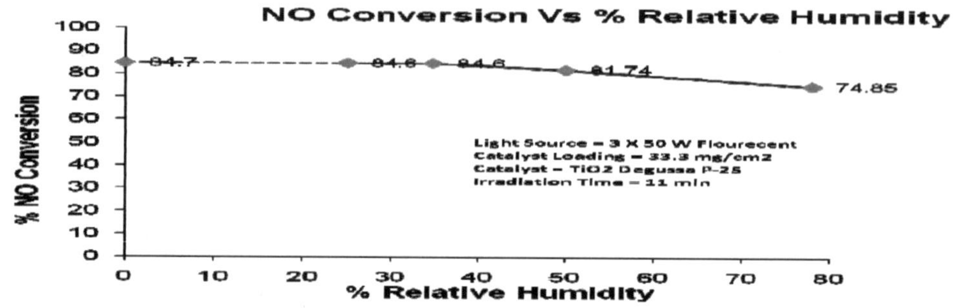

Fig. 3 Effect of relative humidity @ Inlet NO Concentration 1000 ppb, RT=5 min

According to the ACI Mix Design Code, the water cement ratio by weight for non-air entrained concrete for 25MPa compressive strength in 28 days is 0.61. Since CCP is already hydroxylated, no need for any excess amount of water even though water is an essential contributor to the OH radical. As shown in Fig. 3, for 1000 ppb inlet NO and 5 min RT, NO conversion is essentially the same for RH = 0-35%; For RH > 35%, NO conversion drops due to the competitive adsorption between water & NO [17].

Study of NO_2 Conversion versus Irradiation Time

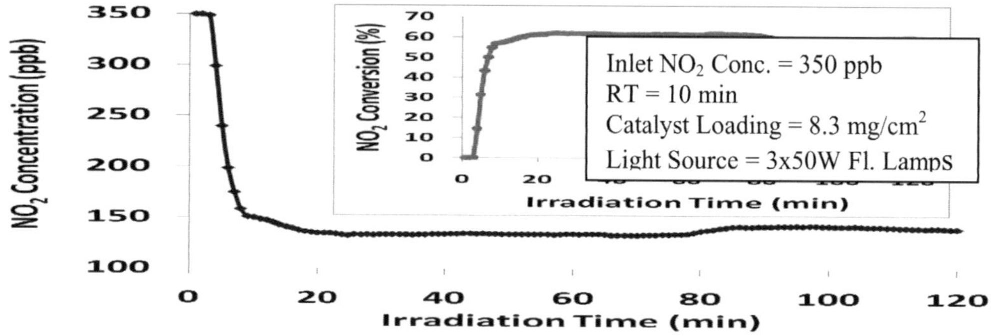

Figure 4 NO_2 Concentration and Conversion Versus Irradiation Time

Figure 4 shows NO_2 concentration profile vs. irradiation time and the figure embedded shows NO conversion exceeding 60% for an inlet NO_2 concentration of 350 ppb with a residence time of 10 minutes. Thus the NO_2 oxidation is slower as compared to NO oxidation. More work needs to be done for other operating conditions

Ion Chromatograph Test For Nitrite and Nitrate Analysis

After a total of 24 experiments were run, the concrete was washed with deionized water. The liquor was tested for NO_2^- and NO_3^- with a Dionex ICS-1000 Ion Chromatograph (IC). The nitrogen mass balance for the adsorbed species is satisfactory from both the analyzer estimate and the IC results (within 10%). The mass balance of NO, NO_2, NO_2^- and NO_3^- confirms the oxidation of NO to HNO_2 to HNO_3 [12,16].

NO Oxidation Products

When the fluorescent lamp is first turned on, the •OH radical quickly reacts with adsorbed NO to form adsorbed HNO_2, as indicated in Figure 5 (a). At this point, the adsorbed HNO_2 can dissociate to H^+ and NO_2^- or react further with •OH radical to form adsorbed NO_2 and water. The next reaction in series, *i.e.*, NO_2 reacting with •OH radical to form HNO_3, is always fast and eventually reaches equilibrium with its reverse reaction on the surface of the catalyst[12]:

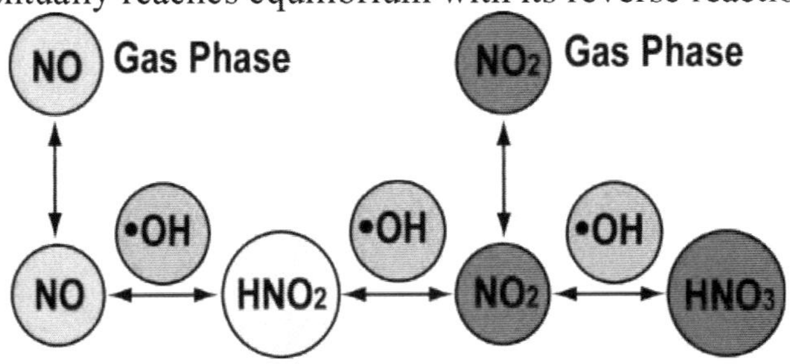

Fig. 5 (a) NO oxidation pathway on catalytic concrete surfaces

Figure 5(b) shows the outlet concentration profile (on a log scale) for an 1000 ppb inlet NO and 5 min and 10 min. RT, which indicates the oxidation of NO to NO_2

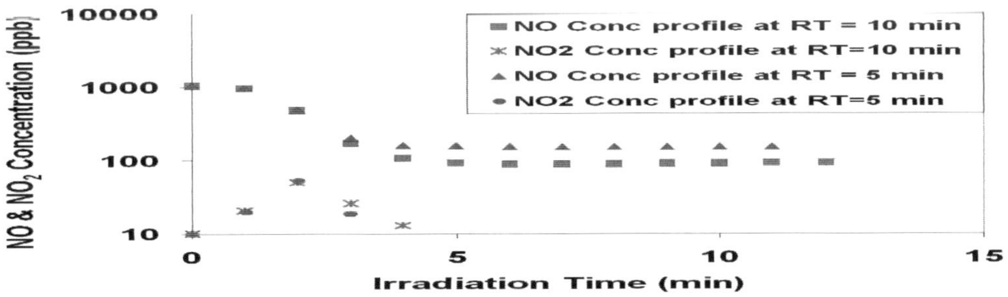

Figure 5(b) NO/NO_2 Concentrations Versus Irradiation Time

[12,18]. The experiment is performed at two different residence times (RT), 5 min and 10 min.

TiO_2 Modified with $BaTiO_3$, Nd_2O_3, and Fe_2O_3 on NO Conversion (Fiber Optic Support)

Figure 6 compares the NO conversion using Degussa P-25 TiO_2 and TiO_2 modified with $BaTiO_3$, Nd_2O_3, and Fe_2O_3. As shown, the steady-state NO conversion for 2 wt% of Fe_2O_3 and 0.15 wt% of Nd_2O_3 (the balance P-25TiO_2) are higher than that of the baseline Degussa P-25 TiO_2. Noticeably NO_2 release to the gas phase is minimal for boosted TiO_2, indicating the enhanced adsorption of NO_2 and oxidation to the final product HNO_3.

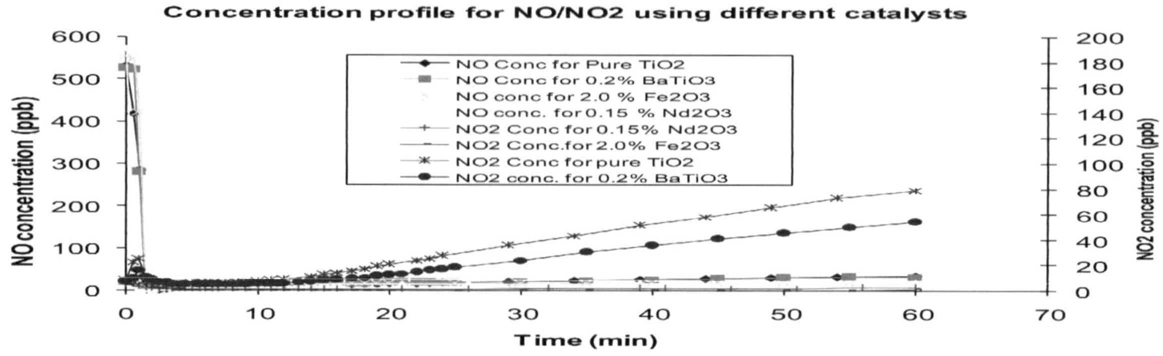

Figure 6 Comparison of NO oxidation for baseline and boosted TiO_2

Concrete Photoreactor with a Uniform Catalytic Layer

Several catalyst-coated concrete photoreactors with small dimensions of 18.5" X 18.5" X 5" were fabricated according to the ACI mix design with a 3-mm catalytic layer with Type 1 Portland cement :Degussa P-25 TiO_2 = 2:1 (weight ratio) and TiO_2 : SiO_2 : CaO : Fe_2O_3 = 1 : 0.58 : 1.77 : 0.09 (mole % ratio). Three 8 watt UVA (Black Light) gives a total radiation exposure is 110 watt/m^2 . The experimental apparatus is essentially the same as in Fig. 7 except more instruments such as ozone analyzer are added. Catalyst loading is 25 mg/m^2 in 3 mm (~1/8") uniform layer. Most experiments were carried out either at 25% relativity humidity (RH) or without adding water vapor (RH=0%). The results indicate that surface coating of TiO_2 is much more effective compared to a uniform layer in terms of NO conversion and NO_2 adsorption. That also suggests a need to investigate different proportions of TiO_2 to cement mixing ratios since a uniform layer is more durable than pure TiO_2 surface coating.

Residence Time Effect (for a uniform layer)

The NO conversion is increasing from 38 – 55% for the residence time 2 – 7 for the inlet NO concentration of 1000 ppb without adding humidity. As the residence time is increasing there is more time for the NO to react photo catalytically. NO_2 yield is defined as $[NO_2]/([NO]_0-[NO])$ where [NO], $[NO_2]$ are NO & NO_2 concentrations, $[NO]_0$ is inlet NO concentration. Both NO_2 concentartion and NO_2 yield decrease with the incresing RT from 2-7 min. Increasing RT allows more time for the oxidation reaction to go to completion (HNO_3).

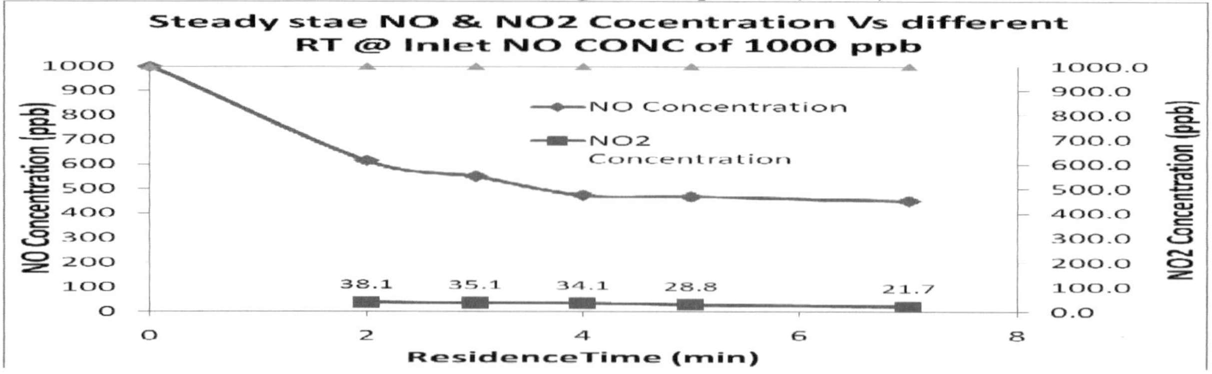

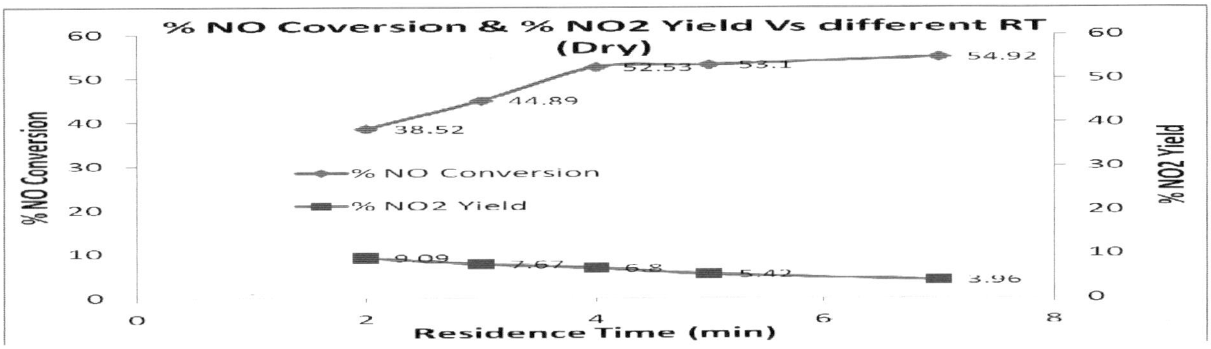

Fig 7: % NO Conversion & NO_2 yield vs Residence Time (without humidity)

Inlet Concentration Effect (for a uniform layer)

The inlet concentration effect were studied @ 5 min Residence time for both dry gases and at 25 % RH. The results are similar. As shown in Fig. 8, the NO conversion decreases from 51 – 28% with increase in inlet NO concentration from 200 – 2000 ppb.

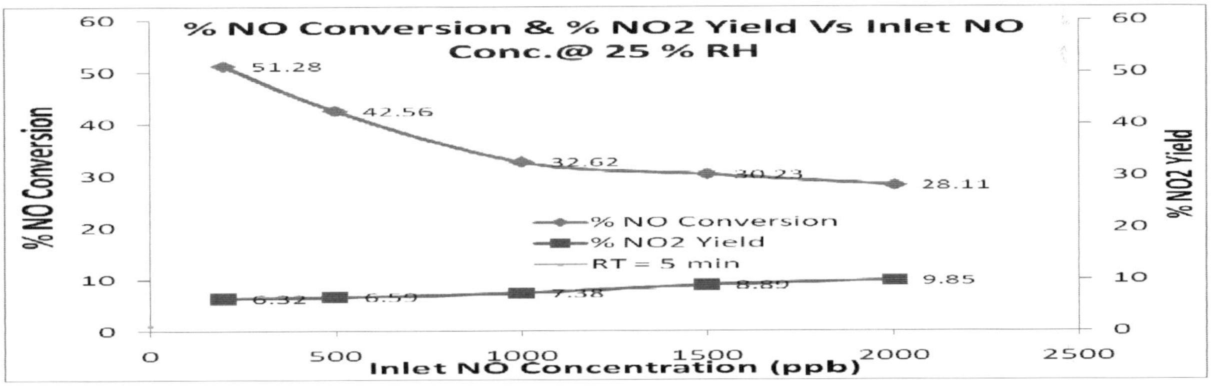

Fig 8: % NO conversion & NO_2 yield vs Inlet NO Concentration for 25% relative humidity

NO convesion for different catalayst uniform layers

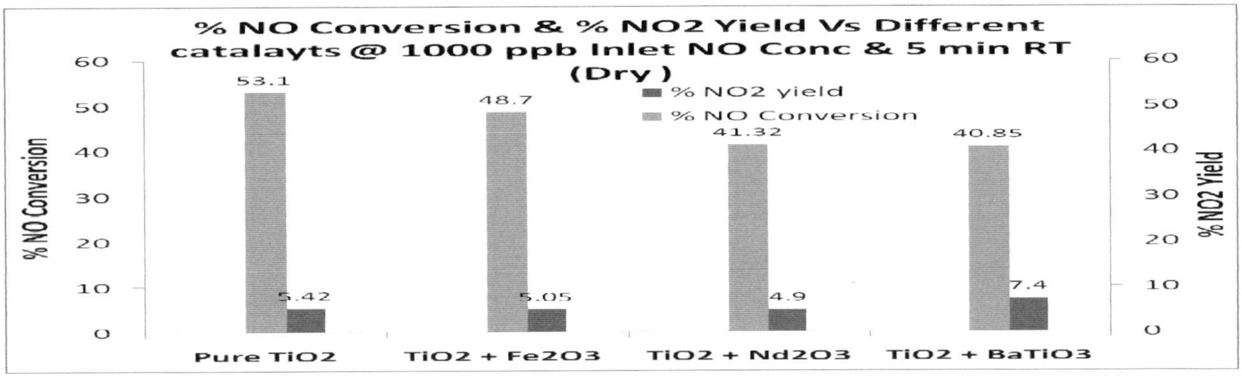

Fig 9: % NO conversion & NO_2 yield vs different types of catalysts Coated on CCP

Optimize TiO_2: SiO_2: CaO:Fe_2O_3 Ratio

Figure 10 shows twin peaks in the photocatalytic activity of a mixed TiO_2/SiO_2 catalyst synthesized with the sol gel technique. The activity is measured with the oxidation of atrazine, an herbicide, in the aqueous phase under fluorescent light. The interesting phenomena about this result is that pure TiO_2 does not show the highest activity but a certain ratio equivalent to 94% and 16% TiO_2 mole ratio give surprisingly high activity. This together with the fact that Fe_2O_3 enhance NO oxidation at an optimal 2wt% on glass and fiber optic support and Type I Portland cement contains 3.3 mole percent of Fe_2O_3 [19] offer an opportunity to test various proportions of TiO_2 to cement ratio to optimize the ratio of TiO_2: SiO_2:CaO:Fe_2O_3. This may indicate a change of pore structure or surface area in the sol gel catalyst but may also indicate some synergy between TiO_2:SiO_2 and TiO_2:Fe_2O_3 in certain ratios.

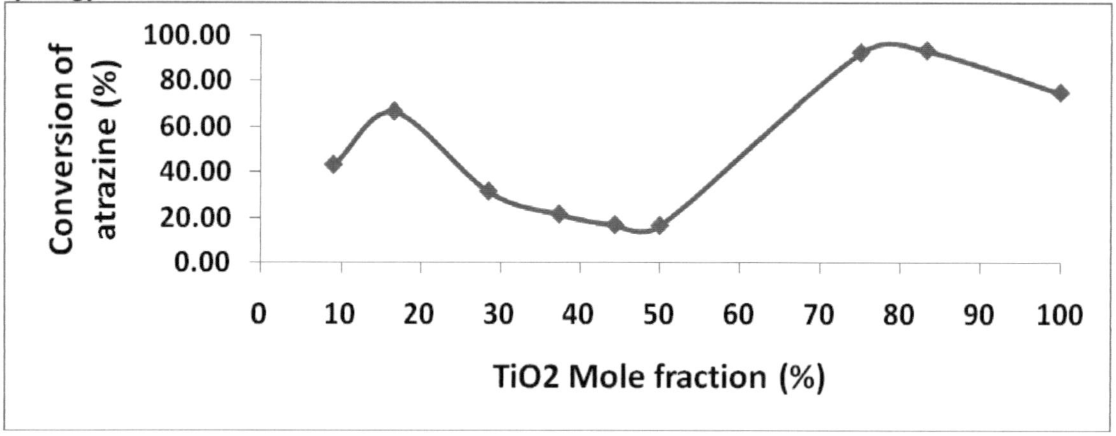

Figure 10 Effect of Mole Fraction of TiO_2 in the TiO_2-SiO_2 Catalyst on Conversion of Atrazine at 60 min Reaction Time

Measurement of O3

Three 15watt UVA lights were used in a 20 L reactor with NO inlet conc.= 1000ppb and RT=5min. No incremental ozone is observed before and after the UVA lamps were turned on or between the inlet and outlet streams during the NO oxidation experiments as shown in Fig. 11. The detection limit is 0.5 ppb. The background O_3 concentrations are between 20-30 ppb.

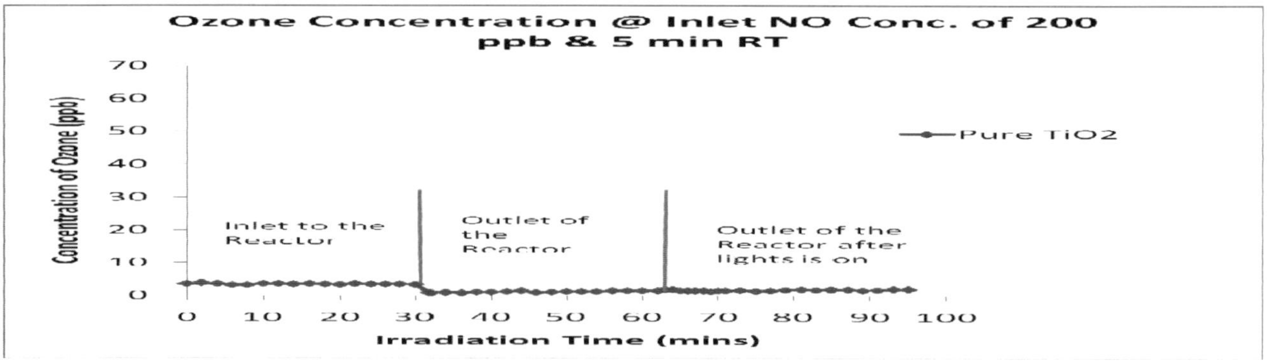

Fig. 11 Ozone measurements @inlet & outlet of the photoreactor coated with P-25 TiO_2

FLUENT Modeling of NOx Plume

In this work, FLUENT 6.3 is used to simulate the NOx dispersion from the exhaust pipe in 3-D. FLUENT follows the finite volume approach to solve the governing transport equations for temperature, pressure, mole fraction and other fluxes [20]. No homogeneous air/NOx reactions or heterogeneous NOx/catalytic surfaces reactions are considered so far but will be included in the third year efforts using the wall surface reactions option in FLUENT. Two conditions have been considered for this study- one is the low-idle condition when the engine runs without the movement of the vehicle for example at a traffic signal or toll-booth and the other is the high-idle condition when the vehicle starts from the stop position. In the low-idle condition, the exhaust exit velocity was taken as 15 m/s and the temperature to be 340 K. In the high-idle condition, the exit velocity was taken as 24m/s and the temperature to be 550K [21]. The composition of the inlet gas was nitric oxide, NO – 350 ppm and nitrogen dioxide, NO_2 – 35 ppm and the background NOx in the atmosphere was NO – 200 ppb, NO_2 – 20 ppb.

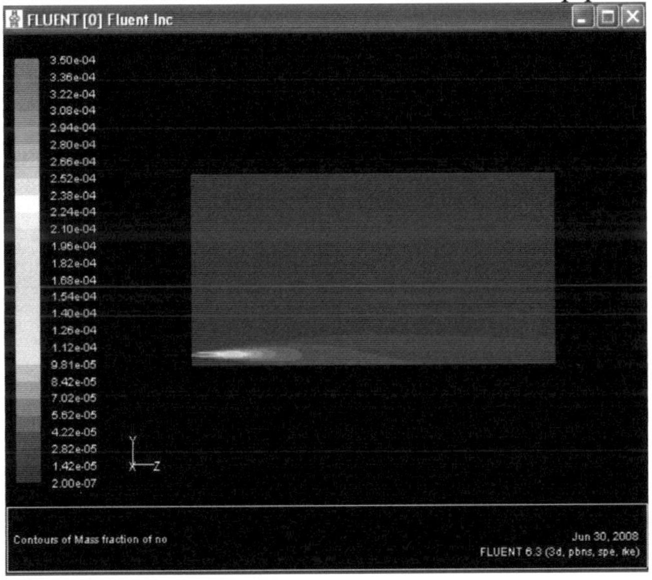

Figure 12 NO Mass Fraction Contours for Low-Idle Condition

The concentration of NOx, the temperature and velocity of the plume from the exhaust pipe decrease along the centerline of the plume and along the distance from the exhaust pipe. The concentration of the NOx falls rapidly by the time plume reaches a distance of approximately 3 m downstream from the exhaust pipe, Fig. 12. The turbulence created between the atmosphere and the exhaust plume helps in dispersing the plume. By the time, the plume travels the distance of 10m along the centerline, NOx disperses and it gets to minimum concentration of 200 ppb in the atmosphere.

ACKNOWLEDGEMENTS
Financial support from USEPA/Houston Advanced Research Center (Grant # 20-23014-TARC122005) and Texas Air Research Center (Grant #077LUB2068A) is gratefully acknowledged.

References

1. http://www.tceq.state.tx.us/subject/subject_air.html
2. EPA/451-K-97-002, Ozone, October 1997
3. EPA 420-R-99-023, Regulatory Impact Analysis, December 22, 1999, IV-13.
4. http://www.epa.gov/air/urbanair/nox/what.html.
5. http://www.epa.gov/air/urbanair/nox/effrt.html

6. National Renewable Energy Laboratory (2001, October). *Photochemical treatment of pollutants* (CDS-SS25-B001). CO.
7. Fujishima, A., Hashimoto, K. & Wuatanabe, T. (1999). *TiO2 photocatalysis fundamentals and applications* (p. 126 – 156). Tokyo: Bkc, Inc.
8. Frantisek Peterka, Photocatalytic surfaces with self cleaning properties, 6th European Commission Conference on Sustaining Europe's Cultural Heritage: from Research to Policy Queen Elizabeth II Conference Centre, London, UK, 1-3 September 2004.
9. EPA Handbook for Advanced Photochemical Oxidation Process, December 1998, EPA/625/R-981004.
10. Hogan, J. 2004. "Smog-busting paint soaks up noxious gases." *New Scientist*. February 4. http://www.newscientist.com/news/news.jsp?id=ns99994636
11. Takeuchi, K. et al."Removal of nitrogen Oxides from Air by Photocatalysis Using Cement-Immobilized Titanium Dioxide," paper presented in *the Second International Conference on TiO2 Photocatalytic Purification and Treatment of Water and Air*, London, Cincinnati, Ohio, October 26-29, 1996.
12. Sid Devahasdin, Chiun-Jr Fan, Kuyen Li, and Daniel H. Chen, "TiO2 Photocatalytic Oxidation Of Nitric Oxide: Transient Behavior And Reaction Kinetics" *Journal of Photochemistry and Photobiology A; Chemistry,* **156**/1-3 pp 161 – 170 (2003).
13. K. Vajifdar, K. Kotnis, B. Ardoin, D. Chen, K. Li, "Photo catalytic Oxidation of PCE over Titania modified by Non Linear Optical Materials", P132, *TOCAT4*, Tokyo, Japan, July 14-19, 2002.
14. K. Vajifdar, T. Han , B. Ardoin, D. Grooms, D. H. Chen, and K. Li, "Fiber-Optic Photo reactor Using Titania Modified with Ferroelectric Optical Crystals for VOC Destruction", *13th International Congress on Catalysis Conference*, Paris, France, July 11-16, 2004.
15. X. Ye, V. Shah, K. Vajifdar, B. Ardoin, D.H. Chen and K. Li, "VOC oxidation with Various Titania Including Visible-Light-Responsive TiO_2 modified with Ferroelectric Optical $LiNbO_3$," *7th World Congress of Chemical Engineering*, Glasgow, Scotland, July 10-14, 2005.
16. K. Vajifdar, D. Chen, J. Gossage, K. Li, X. Ye, G. Gadiyar and B. Ardoin, "Photocatalytic oxidation of PCE and Butyraldehyde over Titania modified with perovskite optical crystal $BaTiO_3$", *Chemical Engineering and Technology*, 30(4), pp 1-8, 2007.
17. H. Tawara, *et. al.*, "Development of evaluation method of air purifying paving blocks for NOx removal capacity", *Proceedings of The Fourth International conference on TiO_2 Photocatalytic Purification and Treatment of Water and Air*, Albuquerque, NM, 24-28, May 1999.
18. Wang, S. M.E.S. Thesis, Lamar University, Beaumont, TX, 1995.
19. http://original.britannica.com/eb/article-9115999/Table-2-Approximate-Composition-of-Portland-Cement
20. Fluent Users Guide, FLUENT 6.3.26; Fluent Users Center, www.fluentusers.com (accessed 06/30/08)
21. Wang, J.S.; Chan, T.L.; Cheung, C.S.; Leung, C.W.; Hung, W.T. Three-Dimensional Pollutant Concentration Dispersion of a Vehicular Exhaust Plume in the Real Atmosphere. *Atmos. Environ.*, **2006**, 40, 484-497.

SUB-MOLTEN SALT ENVIRONMENTALLY BENIGN TECHNOLOGY

Yi ZHANG, Zuo-Hu LI, Hong-Bin XU, Shi-Li ZHENG, Tao QI, Hui-Quan LI, Institute of Process Engineering, Chinese Academy of Sciences, Beijing, China

Introduction

The global industrial growth has spawned environmental concerns on a global scale. As two challenges of the sustainable development, eco-efficiency and ecosystem pressure have widely raised public awareness throughout the world (Jansen, 2003). In some developing countries, the industrialization has stepped into the peak phase of resource consumption. Industrial pollution and eco-environmental injury will become recognized problems impeding the progress of global economy and sustainable development. The key issue in establishing the sustainable economic pattern is the green upgrading of the huge resource-processing systems resulted from process industries (Zhang, 2001). It has become the main goal of global economic revolution to achieve the economic growth with the least resource-environment cost, maintain the harmoniousness of the eco-system, guarantee the effective eco-utilization of natural resources, and build up an evolving ecological economic pattern. Thereby, a highly challenging strategic innovation direction has been pointed out to the scientists at the same time.

Similar with other developing countries, China's mineral processing contributes greatly to its environmental pollution. With the objective of improving resource utilization efficiency and eliminating environmental pollution at the source, an emerging technology, so-called Sub-Molten Salt (SMS) technology, has been developed by the Institute of Process Engineering, Chinese Academy of Sciences. The SMS technology has the potentials to processing many amphoteric metal resources including chromite, bauxite, and ilmenite. In this paper, the definition of SMS media will be interpreted and the advantage of SMS media in hydrometallurgical process will be elaborately discussed. Then the application of the SMS media in manufacturing chromium compounds, as well as the green hydrometallurgical platform on the basis of the SMS media, will be introduced.

Conception of Sub-Molten Salt (SMS) Unconventional Media

Definition of SMS Media

For aqueous electrolyte solution, solution chemistry and thermodynamic investigation covers usually the low concentration range. When we talk to aqueous electrolyte solution, we commonly refer to aqueous electrolyte solution with low concentration. Physical chemistry research on molten salt has become a popular topic for metallurgy and electrochemistry. However, the solution region between molten salt and common electrolyte aqueous solution has rarely been investigated so far due to the difficulty in experimental measurement and theoretical model construction. The application of highly concentrated aqueous electrolyte solution to hydrometallurgical process has achieved a great progress in our work. It was found that in the high concentration region, the physical chemistry properties of electrolyte aqueous solution becomes remarkably different. As shown in Fig. 1, the highly concentrated aqueous electrolyte solution region was called as SMS region in our work.

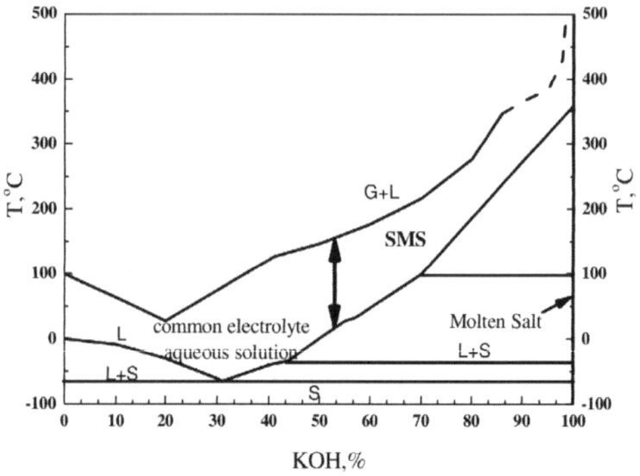

Fig. 1. SMS Region of Aqueous Electrolyte Solution

The SMS unconventional media (Zhang, 2007) was defined as a kind of controllable ionized media of highly concentrated multi-component system with alkali metal hydroxides. It is capable of providing highly active oxygen anion and thereby intensifying reaction and mass transfer process. Taking KOH SMS as an example, in aqueous solution with the presence of oxygen, KOH will be ionized into potassium cation K^+ and hydroxide anion OH^- (Equation (1)). The hydroxide anion OH^- will further be decomposed into water molecule H_2O and oxygen anion O^{2-} (Equation (2)). By combining with oxygen molecule, the oxygen anion O_3^{2-} becomes even more active (Equation (3)).

$$KOH \rightarrow K^+ + OH^- \quad (1)$$

$$2OH^- \rightarrow H_2O + O^{2-} \quad (2)$$

$$O^{2-} + O_2 \rightarrow O_3^{2-} \quad (3)$$

Physical-chemistry Characteristics of SMS Media

The physical-chemistry characteristics of SMS media include the following four aspects. The first is ***high boiling point and low vapor pressure***. For SMS media, a higher solute concentration leads to a higher boiling point, as shown in Fig. 2. Therefore, under the same reaction temperature, the vapor pressure of SMS media is lower than common electrolyte aqueous solution with low concentration. In other words, the SMS media enables a higher reaction temperature under atmospheric pressure.

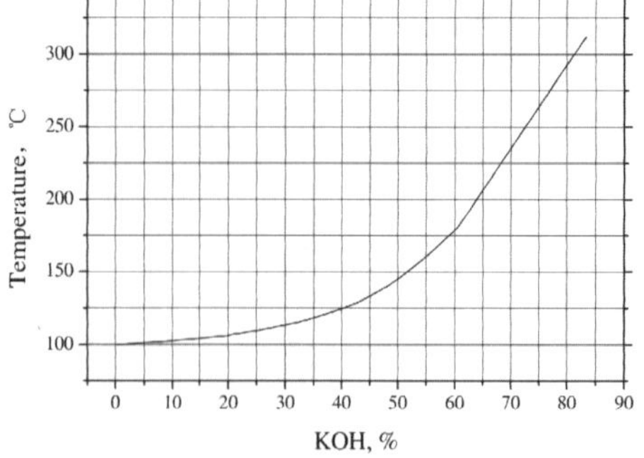

Fig. 2. Relationship of Boiling Point and Concentration for Aqueous KOH Solution

The second characteristic of SMS media is ***high mean ion activity coefficient***. As shown in Fig. 3, in low concentration range, the mean ion activity coefficient increases slightly with the increasing of mole fraction of KOH. However, in high concentration range as well as SMS region, the mean ion activity coefficient increases remarkably with the increasing of mole fraction of KOH.

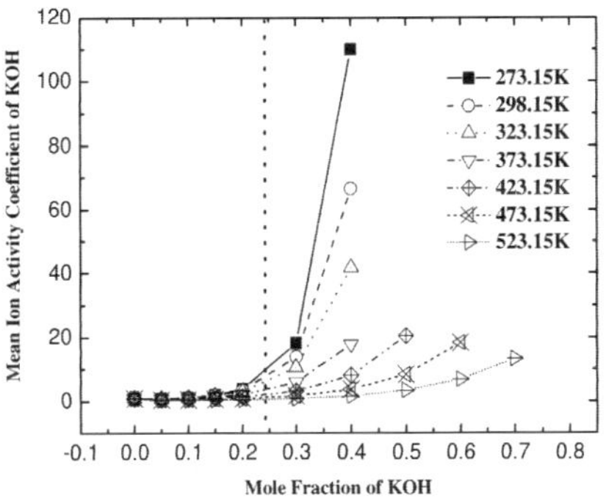

Fig. 3. Relationship of Mean Ion Activity Coefficient and Mole Fraction for Aqueous KOH Solution under Different Temperatures (Data from Li, 1996)

The third characteristic of SMS media is ***excellent performance in flow and transfer***. For SMS media, when operating under higher temperature, its viscosity becomes lower and it will help to improve the fluidity and mass transfer of the aqueous system. The last characteristic is ***easy process control and quantitative adjustment***. As ionic solvent for hydrometallurgical processes, the SMS media achieves the quantitative controlling and system optimization during the process of reaction.

Reaction Intensification Effect of SMS Media

In a typical hydrometallurgical process, some components, including the target component, in mineral ore particles usually become soluble and are dissolved into the aqueous solution system with the interaction of the reaction media under certain temperature, pressure, pH value, etc. The SMS media exhibits many advantages in hydrometallurgical process. The physical-chemistry characteristics mentioned above help to achieve a more environmentally benign process in industrial operation with high reaction temperature but low vapor pressure, excellent fluidity, high reaction activity, and easy control. In fact, the presence of SMS media makes the decomposition of mineral ore particles more efficient and easy. Fig. 4 illustrated the decomposition process of chromite ore in SMS media of KOH.

The high mean ion activity coefficient and high soluble oxygen anion concentration enables to effectively facilitate the transfer of the oxygen anion in the mineral surface, greatly enhance the abnormal change of the mineral surface structure, essentially intensify the decomposition of mineral particles, and remarkably increase the conversion of target component to near 100 percent.

Separation Facilitation Effect of SMS Media

As highly concentrated electrolyte aqueous solution, the SMS media helps to facilitate the separation of target products from the multi-component system due to the salting-out effect of different electrolyte solutes. With reference to the phase equilibrium data and phase diagram, the clean coupling of reaction and separation can be achieved through adjusting the concentrations of different solutes in the aqueous system, as shown in Fig. 5.

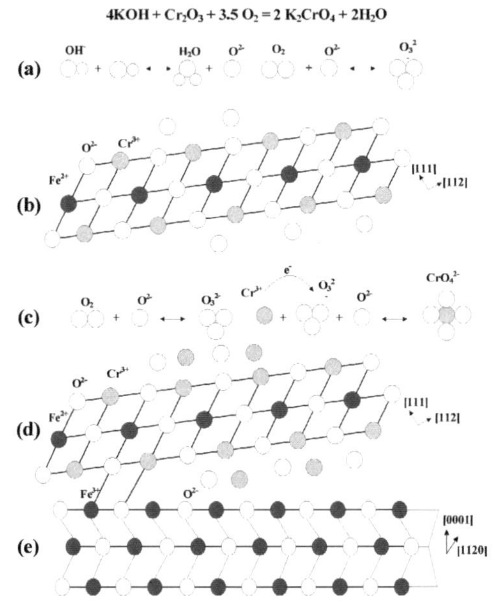

(a) dissolution of oxygen and creation of oxygen anion; (b) main body of chromite ore particle and $FeO{\cdot}Cr_2O_3$ crystal; (c) oxidation of trivalent chromium ion; (d) $FeO{\cdot}Cr_2O_3$ crystal structure destroy, Cr^{3+} dissociation, and product dissociation; (e) creation of Fe_2O_3 product layer.

Fig. 4. Decomposition Process of Chromite Ore in SMS Media

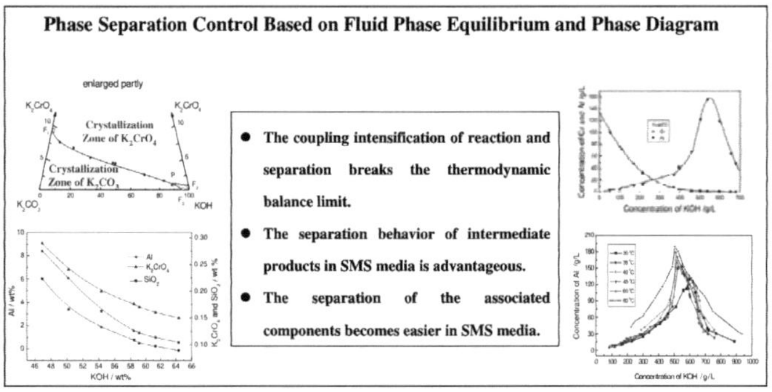

Fig. 5. Phase Separation Behavior of Multiple Components in SMS Media

Case Study: Hydrometallurgical Processing of Chromite with SMS Media

General Background

Chromium compounds are essential to many industries, but their manufacturing process is usually a major source of pollution. In traditional manufacturing process that uses oxidation roasting at high temperature, the utilization efficiency of resources and energy is quite low. Chromate plants discharge large amounts of chromium-containing residues, dusts, and waste gases. Chromium-containing residues create serious pollution problems that threaten groundwater, rivers, and marine areas (Walawska, 2001). By using the SMS media, an environmentally benign hydrometallurgical process of chromite was achieved. The resource utilization efficiency was greatly improved, the energy consumption was remarkably decreased, and the source reduction and zero emission of chromium containing residue was achieved (Zhang, 2005).

Raw Materials Substitution

As illustrated in Fig. 6, the raw materials of the green process with SMS media include only chromite ore, K^+ or Na^+ in form of KOH or NaOH SMS respectively, and air. However, in traditional process, inert additives and sulfuric acid are also employed. The traditional process discharges toxic wastes including Cr-containing residue and Cr-containing Glauber's salt, leading to serious environmental problems. Correspondingly in the green SMS process, the comprehensive utilization of chromite resources is achieved. The ferrite-enriched chromium residue can be used to manufacture desulphurization agent byproduct, and zero emission of Cr-containing residue is achieved.

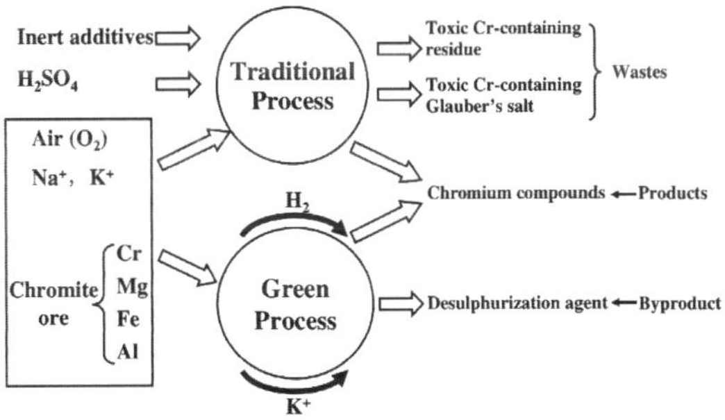

Fig. 6. Raw Materials Substitution in the Green Process with SMS Media

Reaction Mode Innovation

The essence of the green SMS process is that traditional oxidation roasting of chromite ore with sodium carbonate at 1200°C is replaced by continuous liquid-phase oxidation of chromite ore in the KOH SMS media at 300°C in an airlift multiphase reactor, as seen in Fig. 7.

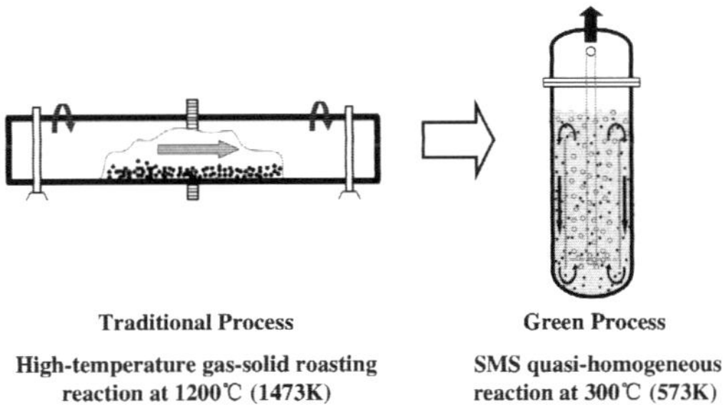

Fig. 7. Reaction Mode Innovation in the Green Process with SMS Media

Technical Indices Comparison

Table 1 showed the technical indices comparison of the green SMS process with the traditional process (Zheng, 2006 and Zhang, 2005). The green process achieved higher extraction yield chromium and zero emission of chromium-containing residue.

Table 1. Technical Indices Comparison of the Green Process with the Traditional Process When Producing 1.0 tonne of $K_2Cr_2O_7$ (in Green Process) or $Na_2Cr_2O_7$ (in Traditional Process)

Item	Green Process	Traditional Process
Chromium recovery yield, %	99	76
Chromium-containing residue, t/t	0.5 (zero emission[1])	2.5
Chromium content in residue, %	0.5	3.0
Hexavalent chromium content in residue, %	0.15	0.7
Chromium-containing waste gas and dust	Almost eliminated	Significant problem
Chromite ore consumption, t	1.07	1.35-1.50
Alkali consumption, t	Below 0.5 (KOH)	0.9 (Na_2CO_3)

Atom Economy Evaluation

Due to the comprehensive utilization of the associated elements such as Al, Mg, and Fe in chromite ore, the atom economy (Trost, 1991) of the green process was much higher than that of the traditional process (Zhang, 2005), as shown in Table 2.

Table 2. Atom Economy Evaluation for the Green Process and the Traditional Process

Item		Green Process	Traditional Process
	Cr, %	98	76
	Fe, %	95	0
	Mg, %	80	0
Atom Utilization	Al, %	90	0
	Na/K, %	98	50
	Ca, %	N/A	0
	S, %	N/A	0
Comprehensive Atom Utilization, %		90.5	14.7

Industrial Scale-up and Demonstration

By use of the green SMS technology described above, a demonstration plant with an annual production capability of 10,000 tonnes of potassium dichromate and 2,000 tonnes of chromic oxide has been built in China (Fig. 8). The technology has exhibited promising results for the industrial production of chromium compounds.

[1] The chromium-containing residue in the green SMS process is used to manufacture desulfurization agent and zero emission is achieved.

(a) (b)

Fig. 8. Photographs of the Chromium Compounds Production Demonstration Plant Constructed in China: (a) Whole Scenery of the Plant; (b) Carbonization Tower.

Construction of Universal Technical Platform with SMS Media

Sketch of the SMS Hydrometallurgical Platform

Extended from the environmentally benign processing of chromite with the SMS media, a universal technical platform for hydrometallurgical processing of mineral resources was constructed and the SMS technology was developed. As shown in Fig. 9, the input material of the universal technical platform is amphoteric metal mineral resources. With the SMS unconventional media, the resource is effectively converted in a green multi-component reaction and separation system. Through an integrated platform for environmentally benign hydrometallurgical process, the objectives of comprehensive utilization of resources and source reduction of pollution can be achieved.

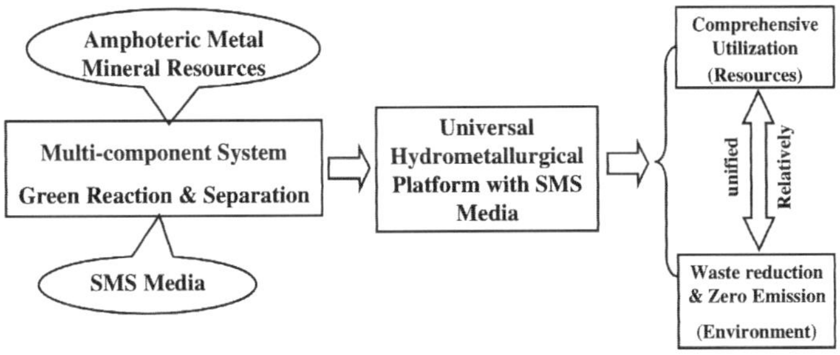

Fig. 9. Sketch of the Universal Hydrometallurgical Platform with SMS Media

Fig. 10 illustrated the target element conversion process for the SMS Platform. The target amphoteric metal oxide minerals are firstly treated with the SMS media and converted into the amphoteric metal oxysalt intermediates, sometimes with the participation of oxygen. The metal oxysalt intermediates are then converted into amphoteric hydroxide hydrate intermediate through hydrolysis or ion exchange process, and sometimes the participation of hydrogen is imperative. By thermal decomposition, the amphoteric hydroxide intermediates can be converted finally into amphoteric oxide products.

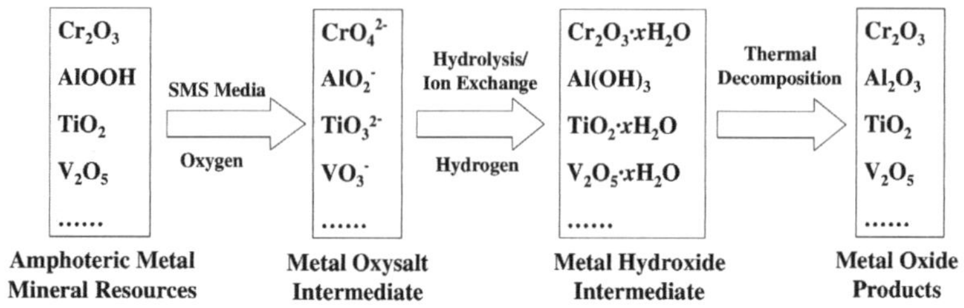

Fig. 10. Target Element Conversion Process for the SMS Platform

Technical Route of the SMS Platform

The SMS platform consists of atom-economic quasi-homogeneous reaction in SMS media, coupling of clean reaction and separation, clean conversion of products, recovery and recycle of media and carriers, and multistage utilization of resource, as can be seen in Fig. 11.

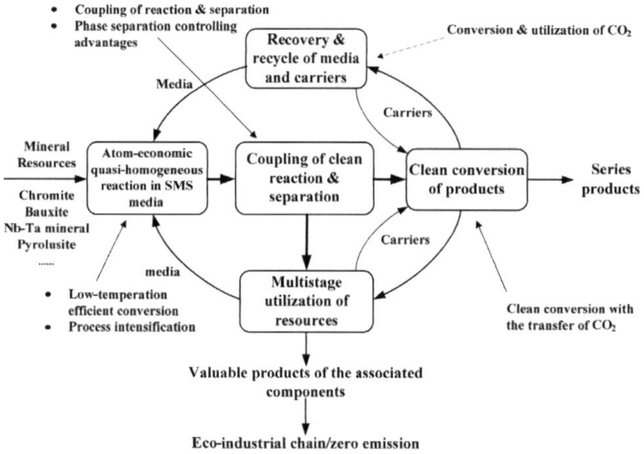

Fig. 11. Technical Route for the SMS Platform (Zhang, 2004)

Applicable Mineral Resources

The universal technical platform with SMS media is generally applicable to varieties of amphoteric metal oxide mineral ore, including Cr, Al, Ti, V, Mn, Nb, Ta, W, Mo, etc. The oxides or hydroxides of these amphoteric metals can react with strong acid as well as strong alkali and produce amphoteric metal oxysalt. The platform has applied to the hydrometallurgical process of chromite, bauxite, and ilmenite. Besides the successful run of the chromium compound demonstration plant, the demonstration plants for manufacturing aluminum oxide and titanium oxide are under construction. The applications of the SMS platform in other amphoteric metals including V, Mn, Nb, Ta have already been investigated in the lab.

Closing Remarks

The SMS unconventional media has excellent physical chemistry properties, which not only simplifies the industrial operation but also helps to intensify the reaction and separation process for the hydrometallurgical processing of mineral resources including Cr, Al, Ti, V, Mn, Nb, Ta, and other amphoteric metals. The universal hydrometallurgical platform with SMS media brings an entirely new approach to environmentally benign hydrometallurgical process. It will effectively resolve the environmental pollution problem in mineral processing industries in such developing countries as China. The application of the SMS platform to the processing of chromite ore has been proven feasible and attractive. The ongoing application of the SMS platform to the processing of other amphoteric metal resources including Al and Ti has exhibited a very promising prospect. The successful development of the SMS environmentally benign technology is revolutionary and will definitely create a new future for hydrometallurgical industries.

Acknowledgements

The authors gratefully acknowledge the financial supports from the National Science and Technology Pillar Program of China (Grant No. 2006BAC02A05), the National Science Foundation of China (Grant No. 50234040), and the National Basic Research Program (973 Program) of China (Grant No. 2007CB613501).

References

1. Jansen, L. (2003). "The Challenge of Sustainable Development", *Journal of Cleaner Production*, 11, pp. 231-245.
2. Li, Z. and Pitzer K S. (1996). "Thermodynamics of Aqueous KOH over the Full Range to Saturation and to 573 K", *Journal of Solution Chemistry*, 25 (9), pp. 814-823.
3. Trost, B. M. (1991). "Atom Economy: A Search for Synthetic Efficiency", *Science*, 254, pp. 1471.
4. Walawska, B. and Kowalski Z. (2001). "Environmental Evaluation of the Effects of Using Chromic Waste in the Production of Chromium Compounds", *Journal of Cleaner Production*, 9, pp. 219-226.
5. Zhang, Y. (2007). "Green Process Engineering for Highly Efficient Conversion and Utilization of Mineral Resources", *The 297th Xiangshan Science Conference*, Beijing, China (March).
6. Zhang, Y. and *et al.* (2005). "Green Manufacturing Process of Chromium Compounds", *Environmental Progress*, 24 (1), pp. 44-50.
7. Zhang, Y. and Chen J. Y. (2004). "Hydrometallurgical Process with Sub-Molten Salt Media", *The Fourth International Conference on Hydrometallurgy*, Xi'an, China (October).
8. Zhang, Y. (2001). "The Green Process Engineering Science", *The Chinese Journal of Process Engineering*, 1 (1), pp. 10-15. (In Chinese).
9. Zheng, S. and *et al.* (2006). "Green Metallurgical Processing of Chromite", *Hydrometallurgy*, 82, pp. 157-163.

Comparative Study of Transesterification of Refined Oil Using a Batch Reactor versus a High Shear Mixer Pilot Scale Reactor

Biodiesel was synthesized with refined soy bean, corn, and canola oil using methanol in the presence of sodium methoxide catalyst. The transesterification reaction was first studied using a 3 liter batch reactor in the laboratory. Following the reactor study a pilot scale process using a 1 gallon per minute high shear mixer reactor was designed and built to mimic transesterification conditions conducted in the laboratory.
In each reactor setting (i.e. 3 liter reactor and high shear mixer reactor), synthesized biodiesel was made in two successive steps with glycerine separation in between. A typical biodiesel synthesis process utilizes an acid to neutralize the sodium methoxide followed by water washing to clean the biodiesel before vacuum drying. Here the biodiesel was cleaned and neutralized using commercially available ion exchange resins (Amberlite BD 10 dry, Dowex DR-G8, and Purolite PD-206). For the first study (3 liter reactor), these resins were mixed in with the biodiesel for 20 minutes. The resins were separated before vacuum drying. In the case of the high shear mixer reactor, the resins were packed into a fixed bed column.

Analysis of both the crude and refined biodiesel in this study included soap content analysis, free fatty acid analysis, as well as free and total glycerine analysis. The total glycerine analysis gave an indication of how well the starting refined oil was converted into biodiesel.

Preliminary experiments were conducted using an IKA labor pilot shear mixer. The configuration of the mixer included a single mixing head generator (stator/rotor). Refined soy bean oil was subjected to a transesterification reaction of triglyceride which was conducted in the presence of methanol solvent and sodium methoxide catalyst. The reaction was divided into two stages. The theoretical conversion for stage 1 was calculated at 90 wt%, whereas the remaining unreacted triglyceride was reacted in stage 2. At the end of the second stage the resulting crude methyl esters were polished using a resin Amberlite BD 10 dry. During the reaction, samples were taken at 15 minute intervals to monitor how the reaction proceeded. Shown below (Figure 1-4) are results obtained from the analysis of the soap content during the reaction and total glycerine.

Based on results in Figure 1-2, high levels of soaps are present throughout the reaction. The presence of these soaps did not seem to impact the conversion of triglyceride. Supplemental results suggested that diglycerides and monoglycerides were formed. Results in Figure 3-4 further emphasize that the reaction was not affected tremendously by the presence of soaps. In addition, Amberlite BD 10 dry removed any impurities present in the crude biodiesel. The levels of free glycerine were minimal. The high soap levels present seemed to favor the glycerine phase during the separation at the end of each stage.

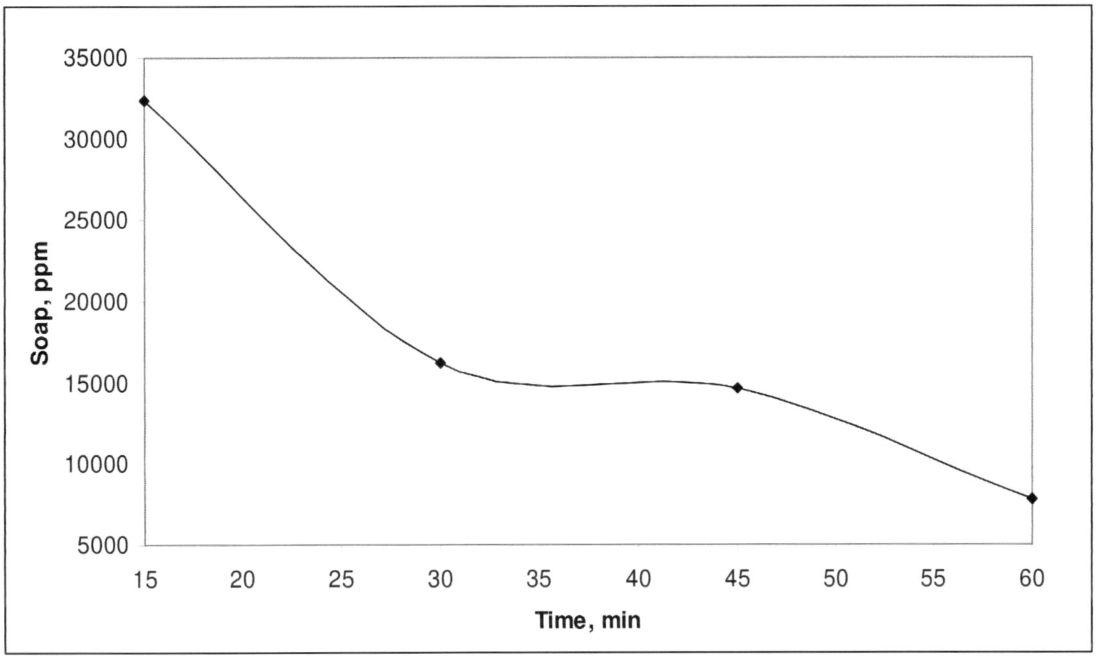

Figure 1. Soap content measurement as sodium oleate during a transesterification reaction Stage 1

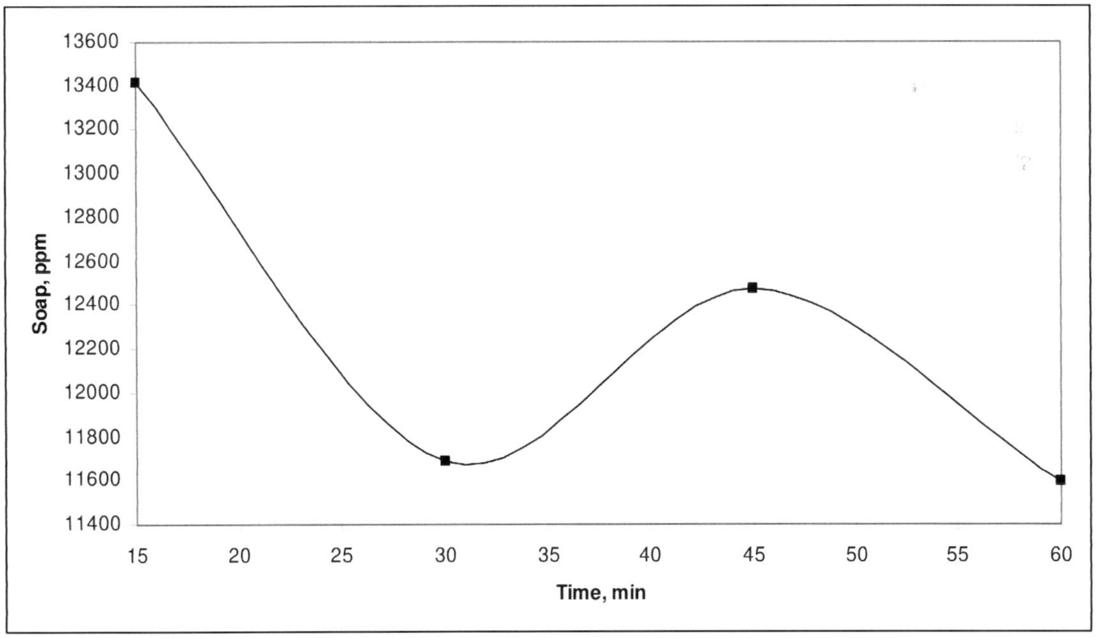

Figure 2. Soap content measurement as sodium oleate during a transesterification reaction Stage 2

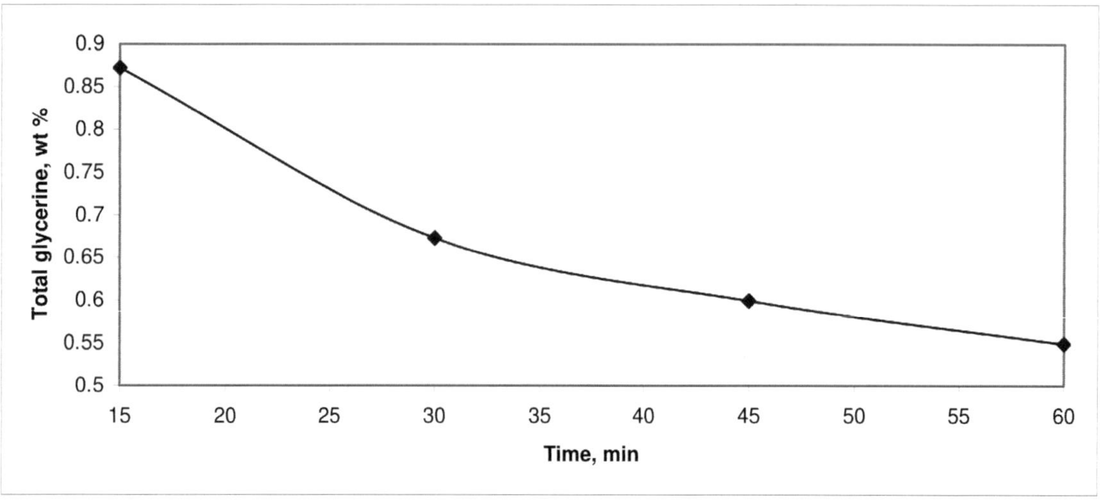

Figure 3. Change in the total glycerine during a stage 1 transesterification reaction

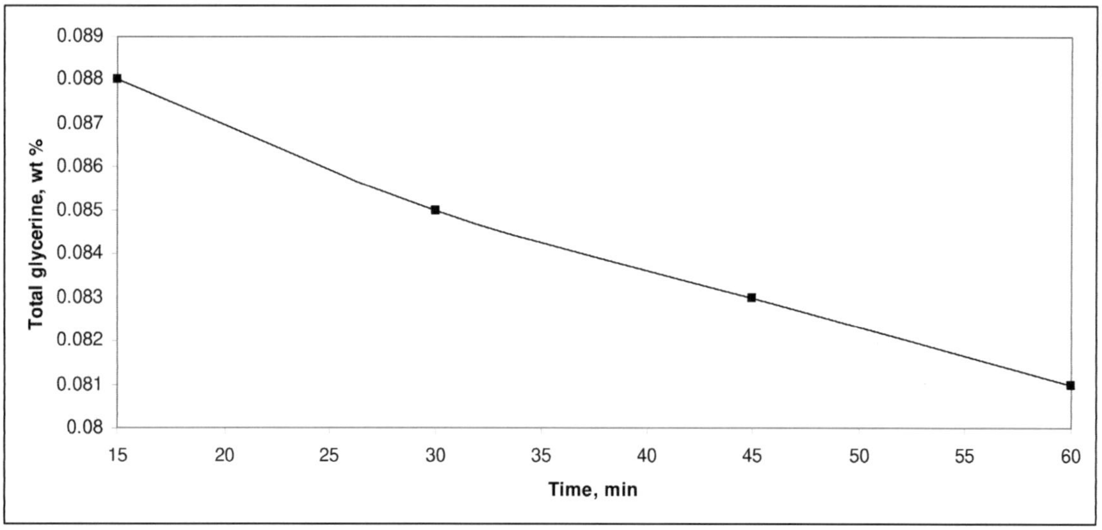

Figure 4. Change in the total glycerine content during a stage 2 transesterification reaction

Economic and Environmental Impact Assessment of Alpha-Naphthol Production by Traditional and Biocataytic Routes

Irvin Osborne-Lee, Gbenga Ajiboye and Jerrad Deason
Chemical Engineering Department, Prairie View A&M University
Post Office Box 519, Mailstop 2505, Prairie View, TX 77446-0519

ABSTRACT

Alpha-naphthol is an important industrial raw material used in the manufacture of insecticides, herbicides, drugs, and dye intermediates. Alpha-naphthol is produced via a Union Carbide patented process, as well as two other chemical predecessor process described in this work. In an effort to find "greener" approach to the conventional synthesis process, the Center of Environmental and Beneficial Catalysis, has proposed the use of a biocatalytic process. The experimental work preformed with computational simulation package have resulted in a process yield approximately 4.5 times higher the compared conventional process and 2.1 yield, with respect to the atom economy . The simulation methods used will also shown an increase in profitability as well as a definite positive environmental impact assessment of the biocatalytic process versus the conventional method.

INTRODUCTION

Reducing pollution has been a major concern leading to the development of a whole new mindset based on "green chemistry", that is, chemistry that is friendly to the environment, minimizes waste, reduces energy utilization and often favors renewable resources over petroleum-based feedstocks. Major goal, for research in the chemical industry is the development of processes and products that have high yield minimizes waste, and are profitable (New Biocatalysts, 2007).

The Center for Environmentally Beneficial Catalysis (CEBC) (Engineering Sustainable Chemicals, 2007) was started to impact and, over time, bring sustainable transformations to the chemical industry that faces many challenges as follows: increasing globalization, societal demands and governmental regulations for higher environmental performance; increased productivity and profitability to maintain global competitiveness; higher consumer expectations of product quality and a looming transition to renewal feedstocks and renewable energy. The perceived long-term outcome of the CEBC and parallel contributors with similar goals is a transformed chemical industry wherein pollution prevention and environmental sustainability replace waste production, pollution, regulation and remediation (The CEBC annual report, year 3 Vol.1).

This work reported here builds on methodology developed within the CEBC for chemical processes. The work then use the methodology to compare a traditional chemical production scheme with a proposed "greener" alternative based on a bio-catalytic route.

OBJECTIVES

In this thesis, several ways to improve environmental performance for chemical processes will be discussed, particularly alpha-naphthol production. These objectives can be achieved by:

a) Developing a baseline scenario for the production of alpha-naphthol based on a traditional chemical route, including process flow sheet, computer simulation, economic and environmental impact analysis.
b) Developing an alternative scenario for the production of alpha-naphthol using a bio-catalytic route, including process flow sheet, computer simulation, economic and environmental impact analysis.
c) Comparing the two alternative scenarios based on performance, cost and environmental impact.

To achieve these set objectives, flow sheets will be developed from technical information obtained from a thorough literature review. VISIO® is the software tool that will be used for the flow sheet representation. Computer simulations will be performed using HYSYS®, in order to obtain the performance specification for the processes described in a and b above. Economic and environmental impact analysis will be performed using SCENE® software.

PRODUCTION OF ALPHA NAPHTOL

There are 37 producers of alpha-naphthol worldwide. The locations of the producers are summarized in Figure 2.1. The figure was developed by the information presented in the directory of world chemical producers-2000 edition. The capacity produced in the western world was approximately over 15,000 tons per year of alpha-naphthol (Trofimov, 1973).

Alpha-naphthol was synthesized by three different ways historically. This includes the following methods.

a) Caustic fusion of naphthalene-1-sulfonic acid.
b) I.G Farbenindustrie process
c) Union Carbide process (Barmicki et.al., 1997).

The most important process to this thesis is the Union Carbide Process, because of its relevance to recent industrial practice. The only difference is in the starting material. Naphthalene is the starting material for the conventional chemical process while tetralin is the starting material for the Union carbide process. The detail description of the conventional chemical process and an alternative biocatalytic process are discussed in further detail elsewhere (Ajiboye 2007).

PROCESS ANALYSIS

Conventional Process

Figure 1 shows a flow diagram for the process of synthesizing alpha-naphthol from naphthalene using the conventional chemical process. Naphthalene enters the hydrogenation reactor at 617°F and 1102 psia in the presence of CoMo-Al_2O_3 catalyst where it is converted to tetralin (Krichko et. al., 1969). Tetralin is then contacted with oxygen entering the oxidation reactor at 188°F and 68 psia where it is converted to tetralin-hydroperoxide (Yamauchi et. al., 1981) as intermediate, which immediately decomposes into alpha-tetralone and alpha-tetralol, and modeled in the decomposition reactor at 122°F and 58 psia. The alpha-tetralone and alpha-tetralol have close boiling points and was separated by fractional distillation. The dehydrogenation reaction takes place in the vapor phase at 716°F, and 40 psia in the presence of a catalyst to produce alpha-naphthol (Kudo et. al., 1976).

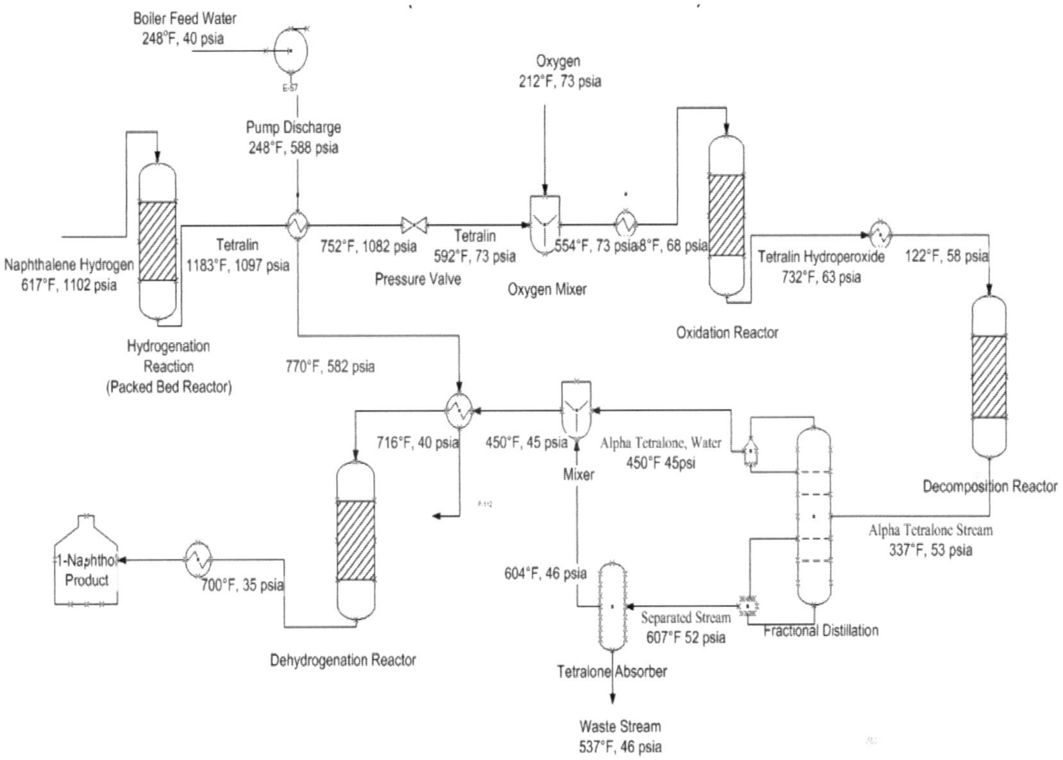

Figure 1 Process flow diagram using Microsoft VISIO®, used as a basis for HYSYS® simulation of the conventional alpha-naphthol production process form naphthalene.

<u>Bio-catalytic Process Development</u>

In 2003, following the challenges indicated by VISION 2020, in a workshop "New Biocatalyst" an alternative approach for alpha-naphthol production, by a greener bio-catalytic method was suggested. The process (see Fig. 2) involved naphthalene oxidation by wild type toluene ortho-monooxygenase with E-coli to perform stereo- and enantiospecific reactions. High selectivity of enzymes reduces the formation of unwanted by-products thereby increasing the atom-economy and efficiency factor (Olivo et.al, 2003). Alpha-naphthol is produced when naphthalene is biodegraded to alpha-naphthol at 30°C; this is a bi-phasic reaction that can be used industrially. (Janardhan, et.al. 2007)

The product is absorbed in both phases and the solvent and organic compounds that contain mainly the makeup stream and naphthalene is recycled back into the feed mixer. This process was chosen as an alternative, because of the promising results obtained by researchers from the University of Iowa, and anticipated to be a good substitute to the conventional chemical process.

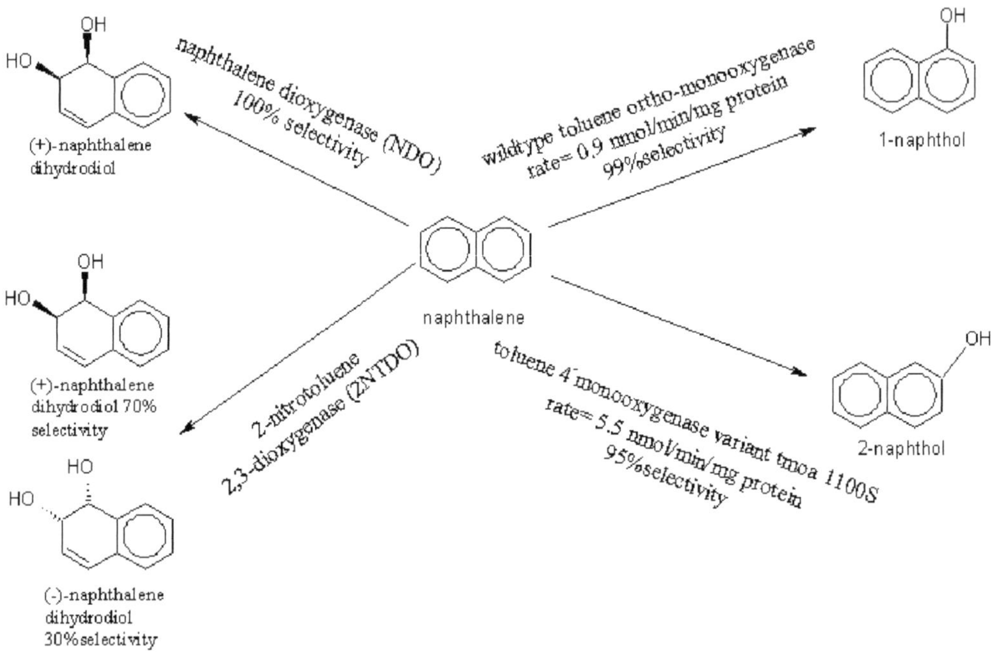

Figure 2. Alternative processes for the production of alpha naphthol. Naphthalene oxidation by wild type toluene ortho-monooxygenase with E-coli (Janardhan et.al., 2007).

Information gathered from the study carried out by researchers at The University of Iowa, on bio-oxidation of naphthalene to alpha-naphthol by the use of ortho-monooxygenase with E-coli, was used in the development of the process flow diagram for the alternative bio-catalytic process as represented on VISIO® by Figure 3.

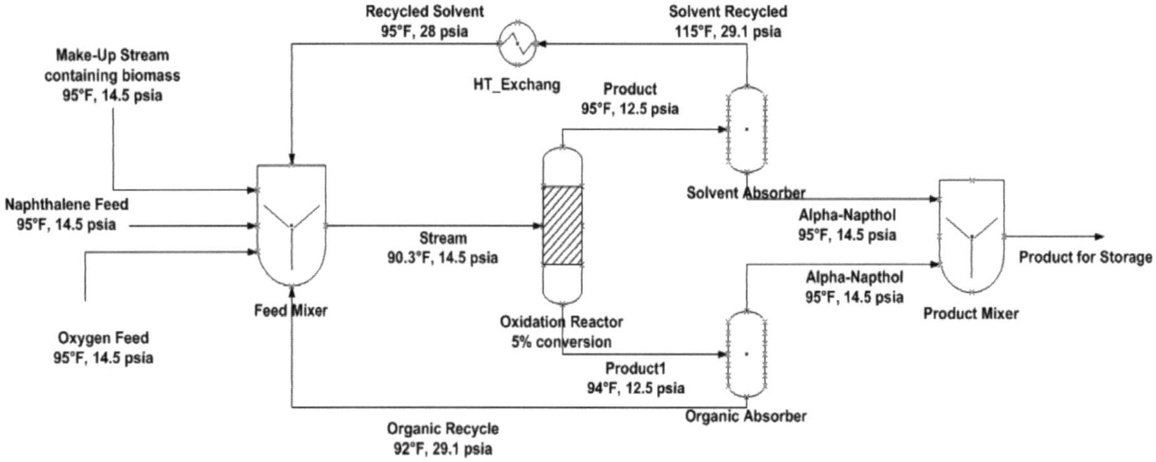

Figure 3. Flow flow diagram of process to be used as a basis for HYSYS® simulation of the alpha-naphthol production process from naphthalene, by bio-catalysis.

PERFORMANCE INDICATORS

The evaluation will be based on comparison of performance indicators 1) yield and 2) atom economy, plus cost and environmental impact. The yield and atom economy are discussed in detail elsewhere (Ajiboye 2007).

COST ESTIMATION

The cost was estimated based on the life cycle cost for a plant, based on major components. The major segments are addressed in the categories of pre-production cost and production cost. The post production costs are assumed to be similar for conventional and alternative processes and hence, are not discussed. Definition of the major elements used in the pre-production cost estimation is discussed in detail elsewhere (Ajiboye 2007).

Environmental Impact Assessment Using SCENE

This study used a simulation software tool called SCENE® to evaluate environmental issues that can be encountered in a manufacturing facilities. The tool introduced nine key environmental indices as explained below. The simulation tool synchronizes with HYSYS® to capture information on the amount of the emissions leaving the process boundary. The chemicals emitted are categorized as to the risk they pose to the environment and the process indices are reported numerically for comparison of process scenarios. Further details of the environmental impact analysis are reported elsewhere (Ajiboye 2007).

RESULTS

Material flows for the two scenarios were determined using HYSYS, which enabled the yield estimates. Similarly, atom economy was determined by material flows for the conventional chemical process and the bio-catalytic alternative. Comparisons of yield and atom economy are presented in Table 1.

Table 1. Comparison of performance Indicators based on the conventional chemical process and alternative bio-catalytic process for synthesizing alpha-naphthol.

Performance Indicators		
	Conventional Chemical Process	**Bio-Catalytic Alternative**
Yield	21.2%	95.7%
Atom Economy	46.5%	100%
Product Mass Flow Alpha-Naphthol	15,000 ton/year	

Life cycle cost was estimated based on the most important elements. Cost for the conventional chemical process and the alternative bio-catalytic process in the categories of pre-production cost, and production cost were determined to produce estimates of plant investment costs, revenue and profitability for values of plant life ranging from 10 to 50 years. These are summarized in Figs. 4 and 5.

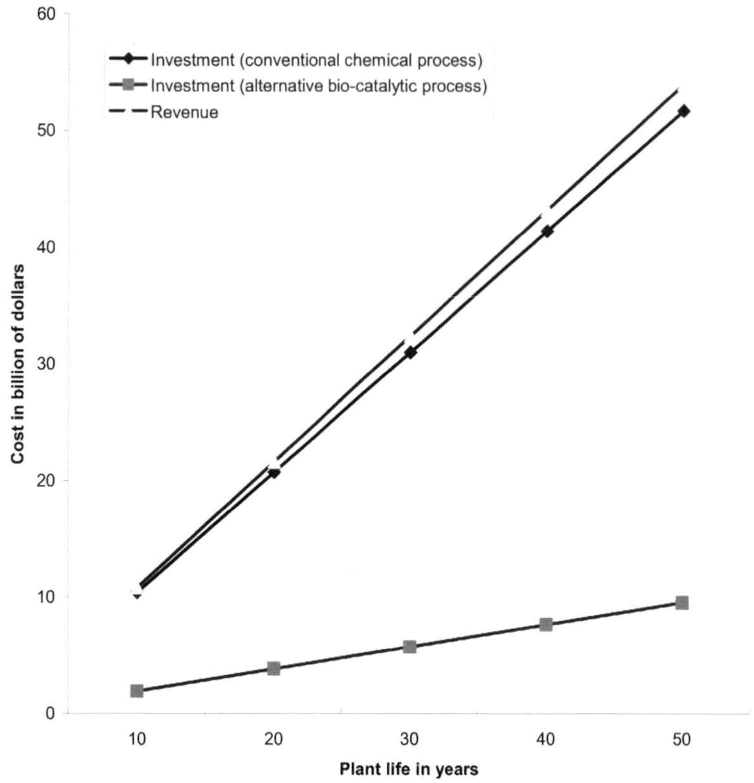

Figure 4 Comparison of variation in total investment and revenue with plant life for conventional chemical process and bio-catalytic process.

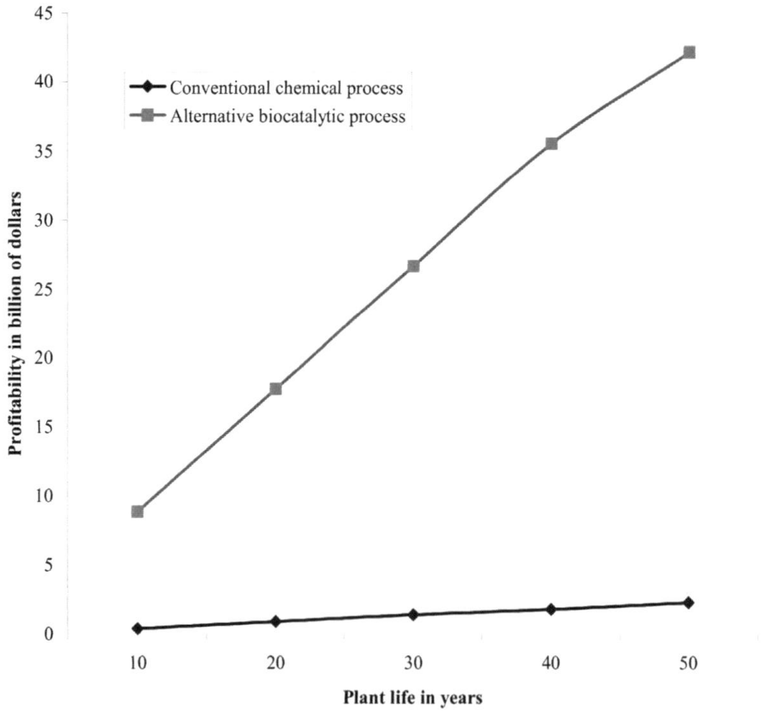

Figure 5. Comparison of profitability with plant life for conventional chemical process and bio-catalytic process.

The environmental impact assessment was made based on results obtained from the simulation tool SCENE.® The results are based on emission of chemicals and risk index summary for the conventional chemical process and the alternative bio-catalytic process. The comparison of the environmental indicators is presented in Fig. 6.

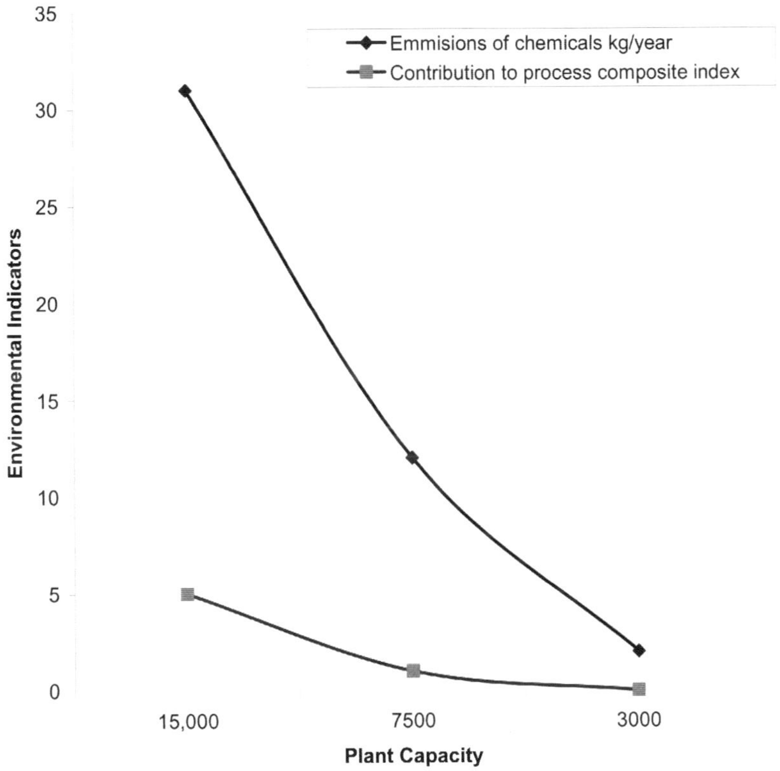

Figure 6. Variation of environmental indicators for conventional chemical process for different plant capacities. Emission of chemicals and risk index summary for the bio-catalytic alternative is negligible (SCENE is only concerned with chemical emission rather than biological waste).

CONCLUSIONS

The need for a greener process development is of recognized importance. This work has made an evaluation of an alternative bio-catalytic process for the synthesis of alpha-naphthol in comparison to a conventional chemical process. The methods used to determine the performance indicators necessary for the comparison use several computer modeling tools such as HYSYS,® CAPCOST,® EPI Suite,® and SCENE® to arrive at results that formed the basis for comparison. Results were presented with comparisons on the basis of performance, life cycle cost and environmental impact assessment.

In term of performance, the conventional chemical process have a yield of 21.2% and an atom economy of 46.5% while in the other hand, the alternative bio-catalytic process have a yield of 95.7% and an atom economy of 100%. In term of yield the alternative process is 74.5% better than the conventional chemical process and 53.5% better in term of atom economy. Based on the result obtained, the alternative bio-catalytic process is a better alternative to the conventional chemical process.

Environmental indicators compared impact based on emission and risk index, it is also very noticeable too that the alternative bio-catalytic process is very green and the waste generated from the process poses no threat to the environment because it is non-toxic, "when it is biological waste is not considered" on the other hand, the conventional chemical process pose a significant threat level. The choice based on environmental impact favors the alternative chemical process because it is greener than the conventional chemical process.

Following the trend of results reported here, the alternative bio-catalytic process appears to be the better choice, because it has a better performance, is more profitable and is greener than the conventional chemical process for the synthesis of alpha naphthol.

REFERENCES

Ajiboye, Gbenga, 2007. "Economic and Environmental Impact Assessment of Alpha-Naphthol Production," M.S. thesis, Prairie View A&M University, Prairie View, Texas.

Barmicki, SD and JJ Siirole, 1997. Separation process synthesis: Kroschita J.I, Howe-Grar M (eds.) Kirk-othmer, Encyclopedia of chemical technology 4th ed, volume 21. John Wiley, New York, pp 923-962.

CEBC, 2006. The Center for Environmental Beneficial Catalysis, Year 3 Annual Report Volume 1.

Directory of world chemical producers, 2000. Manufacturer Listing for a Product Alpha Naphthol. Monday, March 27..

Engineering Sustainable Chemicals and Fuels, the Center for Environmental Beneficial Catalysis, 2007. http://www.cebc.ku.edu/; Visited June 14.

Janardhan, Garikipati, Tonya L. Peeples et al., 2007, Solvent Tolerant Enzyme and Microbial Systems for Aryl Oxidations CEBC Thrust Group1: Catalyst Design, Discovery, and Engineering.

Krichko, A. A. D. V. Skvortsov, T.A Titova et al., 1969. Production of Tetralin By The Hydrogenation of Naphthalene-Containing Fractions, Journal Of Chemistry And Technology Of Fuels And Oils (January), 18-22.

Kudo Et Al., 1976. U.S Patent Number 3,935,282, Jan.27th, Process for the Production of Alpha-Naphthol.

New Bio-Catalysts, Essential tools for a sustainable 21st Century Chemical Industry, 2007. http://www.ccrhq.org/vision/index/roadmaps/New%20Biocatalysts.pdf; Visited, June 14.

Olivo et al., 2003. J. Mol. Cat. B. Enzym. 21. 97-105

Trofimov, V.I., Levkov, Y. and A.M Yakubson, 1973. Sov, Synthesis of Alpha-Naphthol from Tetra-Hydro-Naphthalene Chem. Ind. 3,168.

502c Non-Phosgene Low Temperature Route for Manufacture of Isocyanates

Nitin V. Tople and Eric J. Beckman. Chemical and Petroleum Engineering, University of Pittsburgh, 1249 Benedum Hall, University of Pittsburgh, Pittsburgh, PA 15261

Isocyanates are commonly used as monomers to make polymers, such as polyurethanes which find significant application in the manufacture of rigid and flexible foams. Polyurethanes are also used in the production of adhesives, elastomers, and coatings. Isocyanates are currently manufactured by the phosgenation route which involves reacting an amine with phosgene. This process requires handling of hazardous chemicals and is also environmentally malignant due to issues related to disposal of the hydrochloric acid, formed as a byproduct.

McGhee et al have previously worked on a route to manufacture isocyanates from primary amine, carbon dioxide, a nitrogenous base and an electrophilic dehydrating agent. However, due to the generation of huge amount of salt of the dehydrating agent and the base and due to the absence of an appropriate recycling strategy this route could not be commercialized.

We present a modification to this route by eliminating the use of the nitrogenous base and hence reducing the salt formed. A recycling strategy is proposed to recycle the chemicals used with the goals of making the process commercially viable. A primary amine in presence of low concentrations of carbon dioxide is converted to their corresponding carbamate salt. In presence of a strong dehydrating agent like trifluoroacetic anhydride, the carbamate is converted to isocyanate and salt of the primary amine and trifluoroacetic acid. In excess of carbon dioxide, the zwitterion intermediates of the primary amines are formed which on reaction with a dehydrating agent forms isocyanate without any salt as byproduct. The theoretical isocyanate to amine ratio is 1:2 with the carbamate route and is 1:1 with the zwitterion route. The solubility of carbon dioxide in solvents increases significantly at lower temperatures and hence the zwitterion route is favored in the presence of carbon dioxide excess. Also, results show that the overall conversion is higher at lower temperatures because of the reaction being exothermic. The effect of different solvents and increasingly low temperatures on the overall yield of isocyanates has been studied with the aim of maximizing the isocyanate content formed via the zwitterion route.

McGhee, W.D., Paster M.D., Riley D.P., Ruettimann K.W., Solodar J.A, Waldman T.E, U.S. Patent # 5,451,697, 1995

502d Towards High Purity Separations by a Novel - Sustainable Technology, Fricdiff

Aylin Selvi, André Bardow, and Peter J. Jansens. Process & Energy Laboratory, Delft University of Technology, Leeghwaterstraat 44, 2628 CA, Delft, Netherlands

In the chemical and petrochemical industry, distillation is by far the most important and the most widely applied separation technology and also the largest consumer of energy, especially in the separation of azeotropic mixtures. Besides, the demanding environment conditions stimulate process industry to look for more environmentally friendly and energy efficient production processes. Hence any new separation technology that can lead to important improvements in terms of sustainable development criteria has an incentive to be researched.

Frictional Diffusion, briefly FricDiff, is an alternative separation technology, which emerged from the concepts of reduced energy usage and avoiding hazardous solvents [2].

FricDiff is based on the difference in diffusion rate of two components when diffusing through a third component, so-called counter gas (sweep). In the FricDiff concept, the separation occurs because the light components in the feed mixture diffuse at a higher rate into the counter gas than the heavy components. This difference results into the enrichment of the heavy components on the feed side. The heavy components have a larger molecular weight, and experience therefore more friction with the sweep gas. This friction between the molecules occurs in the pores of a barrier (porous layer), where the feed components and sweep gas diffuse counter-currently.

Previous work dealt with retention of the valuable products either in the feed stream (e.g. isopropanol-water) [2,3] as well as in the sweep stream (e.g. hydrogen- hydrocarbon mixture). For these configurations, the effect of design parameters and operating conditions on the performance of a single tube FricDiff module was studied. A model considering Hagen-Poiseuille flow for feed and sweep side as well as the Binary Friction Model [1] for transport through the porous layer was employed [2,3]. In these studies, only modest enrichments on the order of 65% could be achieved in a single tube FricDiff unit. Such performance is not uncommon for kinetic separation principles but economically not favorable.

In this present work, possibilities to achieve higher purity of the product using FricDiff are therefore investigated. For this purpose, simulation results are analyzed via three different key performance indicators namely enrichment factor, recovery and productivity. Their dependency on the design parameters such as the range of pore size, tube dimensions, the sweep ratio and the flow rate is quantified. The conditions allowing for higher purity separations using FricDiff will be discussed. The simulation results support the claim that the novel FricDiff technology is a sustainable separation method also applicable for tasks of strong enrichment.

Based on these results, a design-case has chosen for a test unit which is under construction now. This unit shall demonstrate the technical viability of FricDiff technology under industrial conditions.

References:

[1] P.J.A.M. Kerkhof, M.A.M. Geboers; Analysis and extension of the theory of multi-component fluid diffusion; Chemical Engineering Science 60 (2005) 3129 – 3167

[2] M.A.M. Geboers, P.J.A.M. Kerkhof, P. Lipman, F. Peters; FricDiff: A novel separation concept. Separation and Purification Technology 56 (1): 47-52 (2007)

[3] A. Selvi, B. Breure, J. Gross, J. de Graauw and P.J. Jansens; Basic parameter study for the separation of a isopropanol–water mixture by using FricDiff technology; Chemical Engineering and Processing, 46 (2007) 810-817.

502e Green Product Design of Temperature Sensors

Maria Nydia Ruiz Felix and Robert P. Hesketh. Chemical Engineering, Rowan University, 201 Mullica Hill Road, Glassboro, NJ 08028

Considerable interest has focused on the design of new products that identify processes, ingredients and systems that are environmental friendly. Chemical products can be manufactured by following different routes. Every route has a planning, research and development cycle. Because of their simple design and low cost, irreversible thermochromic sensors can be used in many applications in the food, textile, and commercial product industries. Initially thermochromic sensors appear white, when the specific temperature is reached, the sensor color changes. This indicates that the substance the sensor was placed on reached the temperature of activation.

Commercial sensors and Rowan sensors were tested using several methods. Parameters such as activation temperatures, activation time, amount of thermochromic mixture in the sensor, mixture quality, and particle size were taken into consideration. The labels were produced and characterized in a temperature range of 61°C to 97°C and 106°C to 131°C. The performance characteristics of Rowan's sensors are similar to the ones currently on the market

502f An Optimal Plant Lay-out Approach Considering Process Safety Issues

Seungho Jung[1], M. Sam Mannan[1], and Richart Vazquez[2]. (1) Chemical Engineering, Texas A&M University, Mary Kay O'Connor Process Safety Center, Artie McFerrin Department of Chemical Engineering, Texas A&M University, College Station, TX 77843-3122, (2) Chemical Engineering, Instituto Tecnológico de Celaya, Av. Tecnológico s/n, Celaya, Gto., Mexico, Mexico

Process plant layout has been an important issue in the chemical industry since it is greatly related to economic profit, saving energy. The objective of the plant layout is to provide the best arrangement of process equipment in a process plant for a given site, and to save more resources such as land occupied by facilities and energy by diminishing piping cost and pump cost.

However, only minimizing the size of the chemical complex to save the economic cost is not the best way for the optimal layout because safety issues have not been adequately addressed. Concern of safety issues, which are not easily accounted for in terms of economic cost, is growing nowadays. At this time, not only the cost of piping, land, and construction but also safety should be included when deciding about the best layout in the field of chemical engineering. Well calculated safety costs, such as losses due to injury and equipment damage, are helpful when obtaining the optimal facility layout ultimately.

This paper presents a new approach for the optimal facility siting considering the possibility of toxic releases, explosions in several facilities. A case study in Beaumont, Texas has been completed to provide a better understanding of how safety cost is assumed and involved in the optimization function. Some occupied buildings, several toxic releases and protection devises were considered here.

The proposed formulation incorporates the effect of wind speed, wind direction and environmental stability to calculate the risk via probit functions and Monte Carlo simulation. The overall problem is initially modeled as a disjunctive program where the coordinates of each facility for siting and cost-related variables are the main unknowns to determine. Then, the convex hull approach is used to reformulate the problem as a Mixed Integer Non Linear Program (MINLP). The analysis results demonstrate the potential of this approach to improve the process design activity.

Web Page: psc.tamu.edu

AUTOMATED TARGETING FOR PROPERTY INTEGRATION

Yin Ling Tan[1], Choon Hock Pau[2], Denny K. S. Ng[3] and Dominic Chwan Yee Foo[3], Raymond R. Tan[4]

(1)School of Engineering and Science, Curtin University of Technology, Curtin University of Technology Sarawak Campus Malaysia, CDT 250, Miri, 98009, Malaysia.
(2)EIS Sampling Sdn. Bhd., No. 148, Lot 3434, 1st Floor, Parkcity Commerce Square, Bintulu, 97000, Malaysia.
(3)School of Chemical and Environmental Engineering, University of Nottingham Malaysia, Jalan Broga, Semenyih, Selangor, 43500, Malaysia.
(4) Chemical Engineering Department, De La Salle University-Manila
2401 Taft Avenue, 1004 Manila, Philippines.

ABSTRACT

Resource conservation is an effective way for reducing operation cost and to maintain business sustainability. Most previous works have been restricted to "chemo-centric" or concentration-based systems where the characterisation of the streams and constraints on the process sinks are described in terms of the concentration of pollutants. However, there are many applications in which stream quality is characterised by physical or chemical properties rather than pollutant concentration. In this work, the automated targeting approach originally developed for the synthesis of composition-based resource conservation network is extended for property-based network. Based on the concept of insight-based targeting approach, the automated targeting technique is formulated as a linear programming (LP) model for which the global optimum is guaranteed. Two literature examples are solved to illustrate the proposed approach.

Keywords: Process integration; Property integration; Resource conservation; Waste reduction; Property interception; Optimisation.

INTRODUCTION

Traditionally, process industries have been focusing on conventional end-of-pipe treatment. However, over the past decades, the center of attention has shifted to more sustainable operations via effective usage of resources. Amongst the few reasons that have resulted in this change include environmental sustainability, stringent emission legislation, as well as increasing of fresh resources and waste treatment costs.

In particular, mass integration which has been applied for material reuse/recycle, has become increasingly important over the past decade. Extensive works and efforts have focused in the area of mass integration. Recent reviews of mass integration can be found in literature (e.g. El-Halwagi, M.M., 1997, 1998, 2006; El-Halwagi and Springs, 1998; Dunn and El-Halwagi, 2003). Mass integration techniques have focused on tracking individual chemical species without considering other properties or functionalities of the streams. In response to this limitation, the notion of property integration, which takes into consideration the properties and functionalities of process stream, has been introduced by El-Halwagi and his co-workers.

Property integration is a functionality-based, holistic approach to the allocation and manipulation of streams and processing units that include the tracking, adjustment, assignment, and matching functionalities throughout the process (El-Halwagi et al., 2004). The concept of property integration is that the problem is mapped to a cluster domain, which achieve the same objective as mass integration (minimum resource usage and waste generation).

To further illustrate the concept of property integration, Shelley and El-Halwagi (2000) introduced a property-based cluster that was driven by tracking functionalities and properties of the streams instead of focusing on the chemical constituents. This technique was aimed to obtain the optimal recovery and allocation of the volatile organic compounds (VOCs) in complex hydrocarbon mixtures. Later, Kazantzi and El-Halwagi (2005) introduced a pinch-based graphical targeting technique, which is generalised from the conventional material reuse/recycle pinch diagram (El-Halwagi et al., 2003). On the other hand, Foo et al. (2006) presented both graphical (property surplus diagram) and algebraic (property cascade analysis) techniques to locate minimum resource targets within a property integration framework. Most recently, Pau (2007) extended the use of property cascade analysis (Foo et al., 2006) to determine rigorous targets for minimum fresh usage, waste discharge and interception targets of resource conservation network (RCN) with interception placement.

It is worth noting that property-based RCN synthesis has been explored in wide range of chemical processes. All the above mentioned works can basically be classified into insight-based and mathematical optimization techniques. To date, some works that utilised both insight-based and mathematical optimisation techniques have also been reported in the area of process integration. Among these works, the automated targeting approach presented by Ng et al. (2008) incorporated the targeting concept of insight-based technique into the mathematical optimisation model to locate the minimum flowrate/cost targets for a RCN. The proposed approach overcomes the limitations when both of the techniques are used independently.

In this work, the automated targeting technique (Ng et al, 2008) is extended for locating the minimum flowrate/cost targets for a property-based RCN. Two literature examples are solved to illustrate the proposed approach.

PROBLEM STATEMENT

Given a problem of resource conservation with single property may be stated as follows: In a process, a set of $N_{sources}$ sources is given, which consist of process streams that may be reused/recycled or discharged. Each source has a flowrate, F_i and is characterised by a property P_i. A set of N_{sinks} process sinks that are process units that can accept sources is also given. Each sink requires a flowrate, F_j and an admissible inlet property, P_j from the source(s), which complies with the predetermined allowable property constraints as follow:

$$P_j^{\min} \leq P_j \leq P_j^{\max} \tag{1}$$

where $P_j^{\min}$ and $P_j^{\max}$ are the specified lower and upper bounds on admissible properties to unit sink j. Besides, a set of N_{fresh} external fresh resources can be purchased to fulfil the requirement of the sink(s).

A general property mixing rule is needed to define all possible mixing patterns among the individual properties. One such form of mixing rule takes the following expression (Shelley and El-Halwagi, 2000):

$$\psi(\overline{p}) = \sum_i x_i \psi(p_i) \tag{2}$$

where $\psi(p_i)$ and $\psi(\overline{p})$ are operators on property p_i and mixture property $\overline{p}$, respectively and x_i is the fractional contribution of stream i of the total mixture flowrate.

The objective of this work is to locate the minimum flowrate/cost targets for a property-based RCN prior to detailed design.

METHODOLOGY

The automated targeting technique was developed by Ng et al. (2008) based on the concept of algebraic targeting technique of cascade analysis (Manan et al., 2004), with the removal of the dual-step procedure. It is worth noting that in all cascade analysis techniques, infeasible cascades with material flow balances are first generated to determine the largest material deficit, which is then added as fresh resource in the second step to remove all deficits and yield a feasible material cascade. Successful application is seen in water network, utility gas network and property-based network synthesis. Via the proposed automated targeting approach, the two-step targeting approach is readily replaced. In this work, the automated targeting is adapted for a property-based RCN.

The procedure for the automated targeting technique for a property-based RCN is next illustrated. A revised *property interval diagram* (Foo et al., 2006) is first constructed, where the property operators (Ψ_k) of the material sinks and sources are arranged in an ascending order, from the lowest level $k = 1$ to the highest level $k = n$ (Figure 1). In cases where the property operator levels for fresh resource(s) and zero property operator level (i.e. 0 MΩ^{-1} (Foo et al., 2006)) do not exist among the process sinks and sources, an additional property operator level is added. Besides, an arbitrarily high property operator level is also added in last level of property interval diagram to allow the calculation of residue property load.

Next, *material flowrate cascading* across all property operator levels is to be performed. At each property operator level k, the difference between the total available material sinks ($\Sigma_j F_{\text{SK}j}$) and sources ($\Sigma_i F_{\text{SR}i}$) may be determined. Equation 3 shows the *net material flowrate* of each k-th level (δ_k). As shown, δ_k is the sum of the net material flowrate cascaded from the earlier property operator level (k – 1), $\delta_{k\text{-}1}$ with the flowrate balance at property operator level k, $(\Sigma_i F_{SRi} - \Sigma_j F_{SKj})_k$, i.e.:

$$\delta_k = \delta_{k\text{-}1} + (\Sigma_i F_{SRi} - \Sigma_j F_{SKj})_k \tag{3}$$

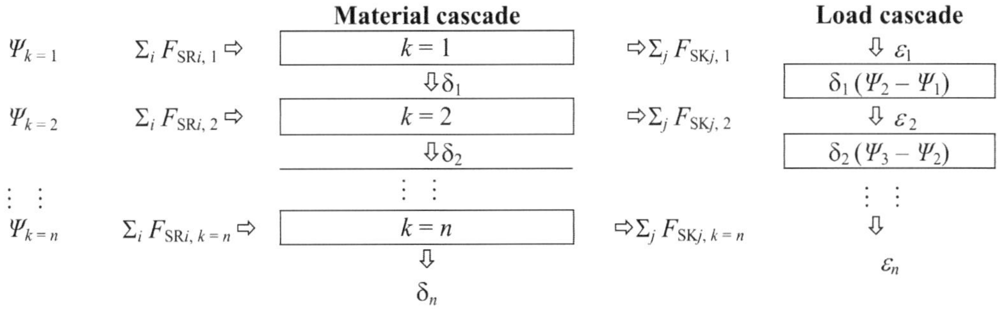

Figure 1 Property interval diagram.

Note that the net material flowrate (δ_k) can either take positive or negative value, with positive value indicates material that flows from the lower level into higher level or vice versa. This is in agreement with the cascade analysis technique.

Apart from material flowrate cascading, *property load cascade* is also essential to ensure a feasible RCN. Property load cascading from level k – 1 to level k is performed as follows. Within each property operator interval, the property load is given by the product of the net material flowrate from level k and the difference between two adjacent property operator levels. Similar to the material flowrate cascade, residual property load of each concentration level k (ε_k) is to be cascaded down to the next property operator level. Hence, property load balance at the k-th concentration level is determined by

$$\varepsilon_k = \varepsilon_{k-1} + \delta_k (C_{k+1} - C_k) \tag{4}$$

where ε_{k-1} is the residual property load that is cascaded from concentration level k – 1.

In order to ensure maximum allowable property load of sink in each level is fulfilled, and the property load is transferred from lower to higher level, the residual property load, ε_k must take a positive value (Equation 5). Therefore, Equation 5 is included as a constraint in the optimisation model. Note also that, a pinch point is observed where the residue property load is zero along the cascade.

$$\varepsilon_k \geq 0 \tag{5}$$

In order to target the minimum fresh resource and waste generation flowrates for a direct reuse/recycle scheme, the property interval diagram may be converted to a *property-based cascade diagram* (PRCD) as in Figure 2. Based on the previous proposed procedure, the property operators of the sink and source are first arranged in an ascending order as shown in Figure 2. Next, utilising Equation 3, the net material flowrate of each property operator level k (δ_k) is determined. It is interesting to note that the net material flowrate found before the first property operator (δ_0) and final property operator (δ_k) levels corresponds to the fresh resource (F_{FR}) and waste discharge (F_W) flowrates of the network respectively (Figure 2). Normally, the fresh resource is of the highest quality, and corresponds to zero property operator. In cases where the fresh resource does not exist at the zero operator value, a new property operator level ($\Psi_{FR\ II}$) is to be added (see Figure 2).

In order to determine the minimum total fresh resource flowrate $\sum(F_{FR} + F_{FR\ II})$, the objective function for the mathematical optimisation model is formulated as:

$$\min \sum(F_{FR} + F_{FR\ II}) \tag{6}$$

subject to the constraints in Equations 3 to 5.

It is worth noting that the above optimisation model is a linear programming (LP) problem, which can be solved easily to achieve global optimal solution if one exists. Two different case studies has been modeled and applied in LINGO software to prove the applicability of the methodology. These case studies have been used to demonstrate the flexibility of the automated targeting approach to analyse the interaction between the pre-treatment systems with the RCN.

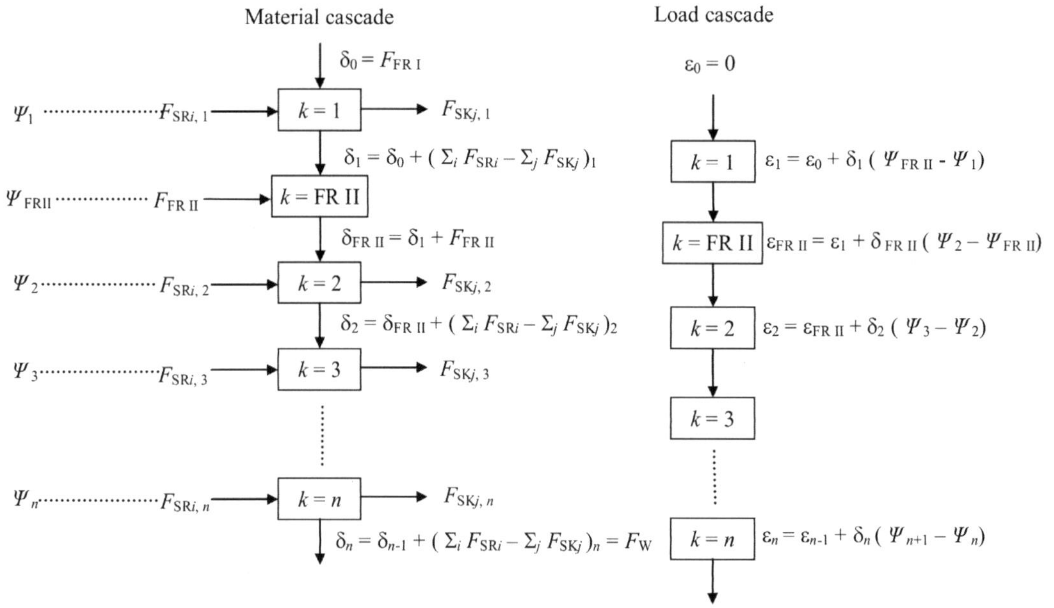

Figure 2 Generic PRCD for direct reuse/recycle.

EXAMPLE 1

A metal degreasing process taken from Kazantzi and El-Halwagi (2005) is used to illustrate the proposed approach, with the schematic flowsheet shows in Figure 3. As shown, a fresh organic solvent is used in the degreaser where reactive thermal processing is used to decompose grease and organic additives. Then, the liquid solvent is regenerated from the solvent regeneration unit and reused in the degreaser; while the gases containing solvent is passed through a condenser and absorber before being flared. Besides, the fresh solvent is used in the absorber to capture light gases that escape from the solvent regeneration unit. Note that the process produces two condensate streams that may serve as process sources, i.e. from the solvent regeneration unit (SR1) and the degreaser (SR2). Reuse/recycle of the process sources (SR1 and SR2) can be considered to reduce the consumption of the fresh solvent. In this case, there are two process sinks where solvent is needed, i.e. degreaser (SK1) and absorber (SK2).

The main property of the solvent that dictates the extent of reuse/recycle of the process sources is the Reid Vapour Pressure (RVP), which is important in characterising the volatility (and, indirectly, the composition) of the solvent. The general mixing rule for the RVP is given as below (Kazantzi and El-Halwagi, 2005):

$$\overline{\mathrm{RVP}}^{1.44} = \sum_i x_i \mathrm{RVP}_i^{1.44} \tag{7}$$

The property operator for RVP, $\Psi(\mathrm{RVP})$, can thus be expressed as follows:

$$\Psi\left(\mathrm{RVP}_i\right) = \mathrm{RVP}_i^{1.44} \tag{8}$$

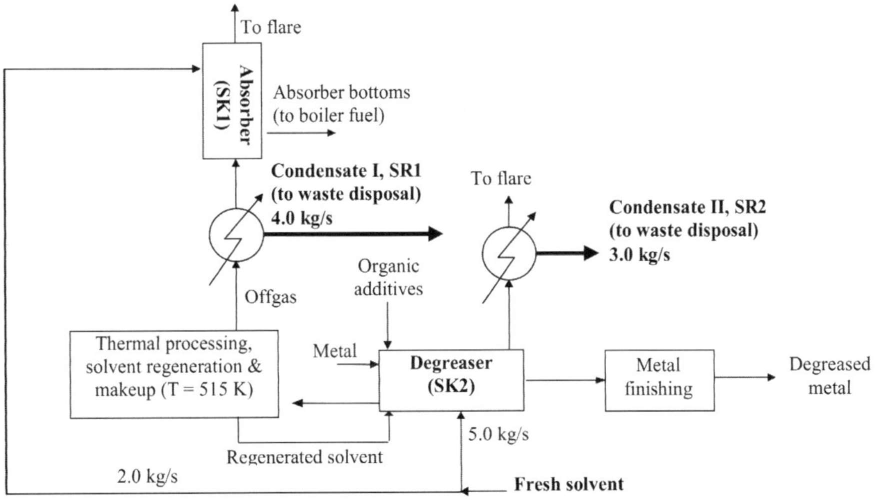

Figure 3 Schematic flowsheet of metal degreasing process (Kazantzi and El-Halwagi, 2005).

Table 1 shows the limiting data for Example 1. Note that the fresh solvent has a relatively low operator value as compare to the two process sources (SR1 and SR2). In order to maximise solvent recovery from the process sources, the upper bound of the operator for each process sink is used to construct PRCD.

Table 1 Limiting data for Example 1 (Kazantzi and El-Halwagi, 2005).

Process	Flowrate (kg/s)	Reid Vapor Pressure, *RVP* (atm) Lower bound	Reid Vapor Pressure, *RVP* (atm) Upper bound	Property operator, Ψ ($atm^{1.44}$) Lower bound	Property operator, Ψ ($atm^{1.44}$) Upper bound
(Sink)					
Degreaser (SK1)	5.0	2.0	3.0	2.71	4.87
Absorber (SK2)	2.0	2.0	4.0	2.71	7.36
(Source)					
Condensate I (SR1)	4.0	6.0		13.2	
Condensate II (SR2)	3.0	2.5		3.74	
Fresh solvent	To be determined	2.0		2.71	

Equation 6 is solved subject to the constraints in Equations 3 – 5, yielding the solution as in the PRCD (Figure 4). Note that, a zero property operator ($\Psi_1 = 0$ $atm^{1.44}$) is added in the first level of PRCD because none of the sink or source with zero property operator. On the other hand, an imaginary property operator level ($\Psi_7 = 100$ $atm^{1.44}$) is added at the last level of PRCD to allow residual property load to be computed. As shown in Figure 4, the targeted fresh (F_{FR}) and waste solvent (F_W) flowrates are both found as 2.38 kg/s, matching the reported results in the literature (Bandyopadhyay, 2006; Kazantzi and El-Halwagi, 2005; Foo at al., 2006). The pinch point is identified as 13.2 $atm^{1.44}$, where zero residual property load ($\varepsilon_5 = 0$) is observed. The network design for this case is shown in Figure 5.

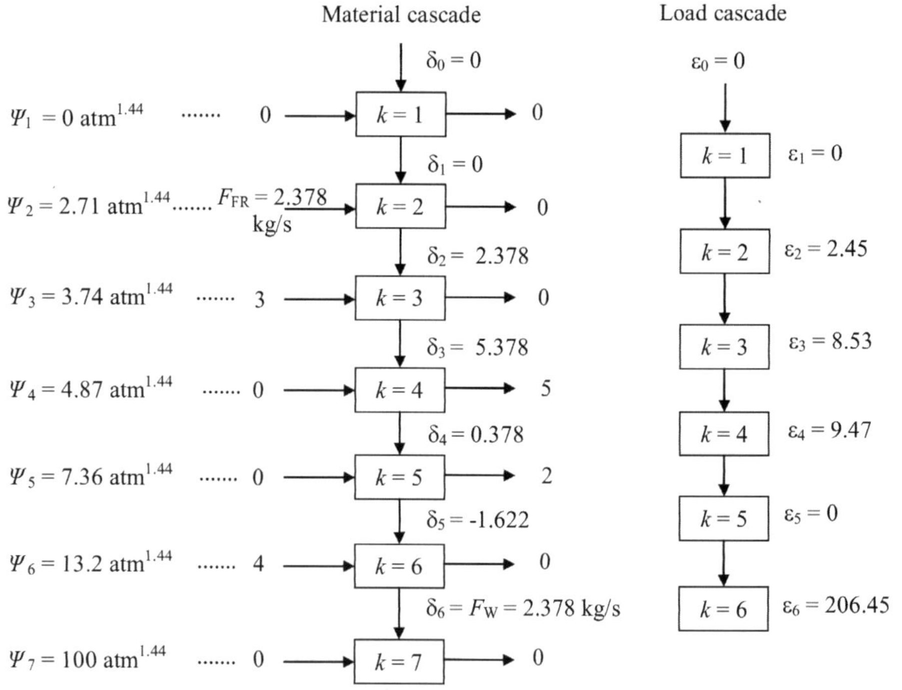

Figure 4 PRCD for Example 1.

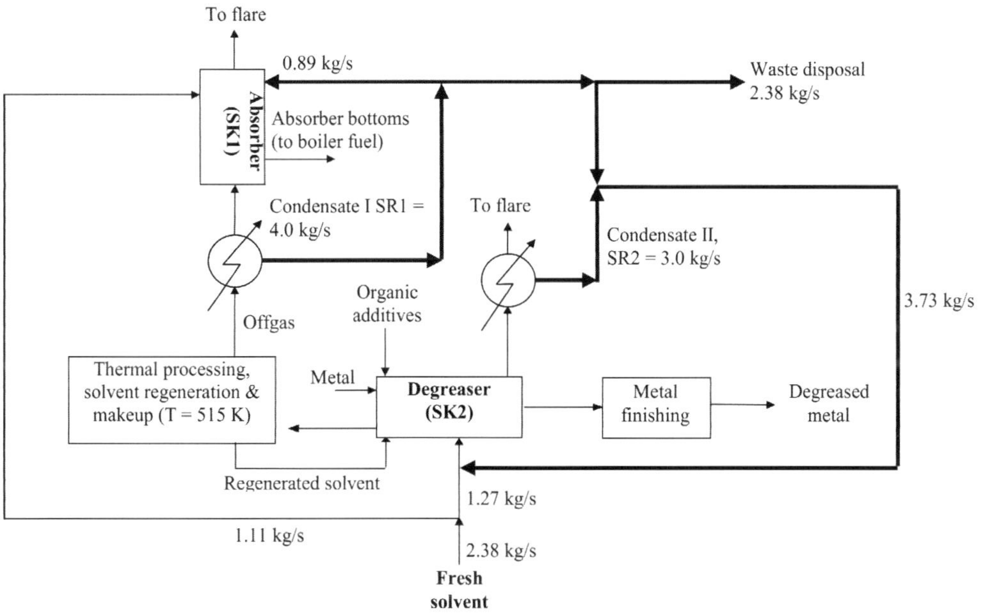

Figure 5 Network design for metal degreasing process with solvent reuse/recycle (Example 1).

EXAMPLE 2

An industrial wafer fabrication process is used to illustrate the competence of automated targeting for the synthesis of a RCN with pre-treatment system. In this example, the pre-treatment system is used to generate ultra-pure water to fulfil the process requirements. The water pre-treatment system consists of three main elements which are ultra-filtration (UF), reverse osmosis (RO) and deionisation (DI). The ultra-filtration is used to retain the solute of high molecular weight in the municipal fresh water. Meanwhile, reverse osmosis (RO) and deionisation (DI) processes are used to remove the ions that solute in the water and deionise the water in order to generate ultra pure water (UPW). In this work, it is assumed that 70% of the inlet flowrate of UF passes through the membrane as permeate; while, 30% of the flowrate is rejected as wastewater with constant water quality, and the same assumption applies for the RO in the pre-treatment system. In order to reduce the fresh water consumption, the recovery of rejected stream from the pre-treatment system can be considered. It is notable that the rejected flowrate is dependent on the amount of water that is fed to the pre-treatment system. This dependency enables the interactions between pre-treatment system and water network to be explored.

Figure 6 shows a schematic diagram for wafer fabrication process with water pre-treatment system. As shown in Figure 6, there are four sections in the wafer fabrication (FAB) process that require UPW supply, denoted as "Wet," "Lithography," "CMP" (i.e., combined chemical and mechanical processing) and "etc." (other miscellaneous processes). Besides, cleaning section, cooling tower makeup and scrubber as well as pre-treatment system necessitate an external supply of municipal fresh water. Note that the Wet and CMP sections generate two wastewater streams with different water quality levels. Besides, there are seven water sources that can be considered for water recovery. In this example, the most significant water quality factor was determined to be resistivity (R) which constitutes an index of the total ionic content of aqueous streams. The general mixing rule for resistivity is shown as below (El-Halwagi, 2006):

$$\frac{1}{\bar{R}} = \sum_i \frac{x_i}{R_i} \tag{9}$$

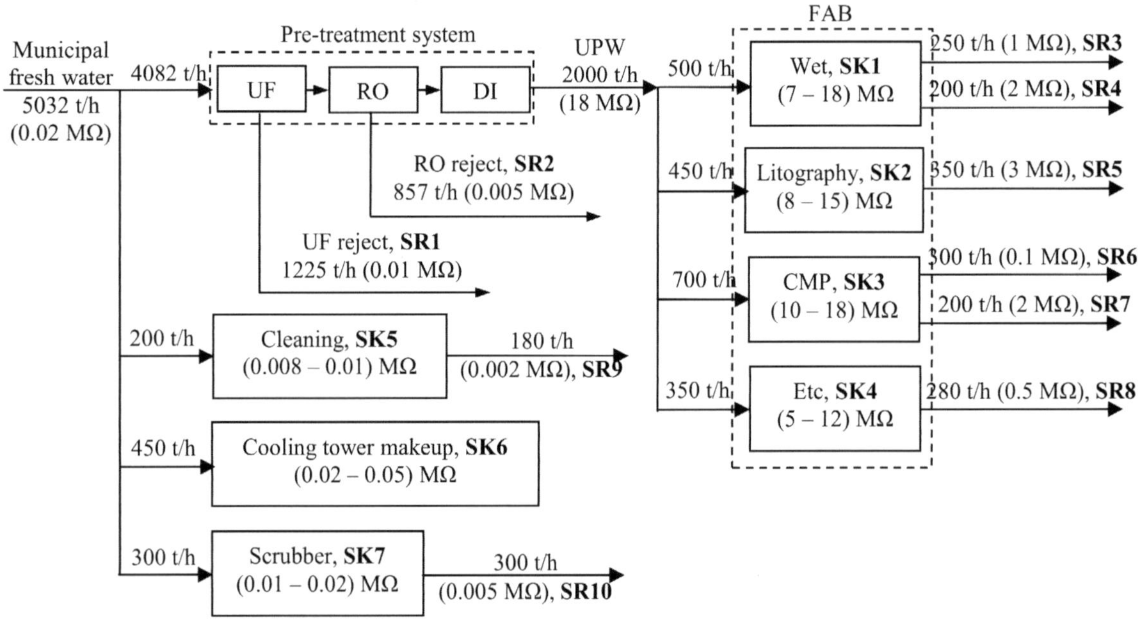

Figure 6 Schematic diagram for wafer fabrication process (Example 2).

Note that the property operator for resistivity is defined as the inverse of resistivity (R^{-1}), such that the lowest resistivity also corresponds to the lowest quality level. Table 2 summarises the pertinent data for the sinks and sources. In this example, the lower bounds of the resistivity are selected as the limiting property for process sink when water recovery scheme is considered, because these lower bounds correspond to the lowest stream quality that can be tolerated by the processes, and thus maximises the potential for reuse and recycling.

Since the rejected wastewater from UF and RO systems depend on the inlet flowrate of municipal fresh water into pre-treatment system, equations below are also included in the optimisation model.

$$F_{\text{In}}^{\text{UF}} = F_{\text{P}}^{\text{UF}} + F_{\text{R}}^{\text{UF}} \tag{10}$$

$$F_{\text{In}}^{\text{RO}} = F_{\text{P}}^{\text{RO}} + F_{\text{R}}^{\text{RO}} \tag{11}$$

where $F_{\text{In}}^{\text{UF}}$ and $F_{\text{In}}^{\text{RO}}$ are the inlet flowrate to the UF and RO system respectively. Meanwhile, F_{P}^{UF} and F_{R}^{UF} represent permeate and retentate flowrates for UF system; F_{P}^{RO}

and F_{R}^{RO} are the flowrate of permeate and retentate streams for RO system. Since all the permeate flowrate of UF and RO systems is directly sent to next treatment unit to generate UPW; therefore, F_{In}^{RO} is equivalent to F_{P}^{UF}, and UPW flowrate (F_{UPW}) is same as F_{P}^{RO}. By combining Equations 10 and 11 leads to equation below:

$$F_{In}^{UF} = F_{R}^{UF} + F_{R}^{RO} + F_{UPW} \tag{12}$$

Table 2 Limiting data for Example 2.

Process	Flowrate (t/h)	Resistivity, *R* (MΩ) Lower bound	Resistivity, *R* (MΩ) Upper bound	Property operator, Ψ (MΩ^{-1}) Lower bound	Property operator, Ψ (MΩ^{-1}) Upper bound
(Sink)					
Wet (SK1)	500	7	18	0.1429	0.0556
Litography (SK2)	450	8	15	0.1250	0.0667
CMP (SK3)	700	10	18	0.1000	0.0556
Etc (SK4)	350	5	12	0.2000	0.0833
Cleaning (SK5)	200	0.008	0.01	125	100
Cooling tower makeup (SK6)	450	0.02	0.05	50	20
Scrubber (SK7)	300	0.01	0.02	100	50
(Source)					
UF reject (SR1)	30% inlet flowrate of UF	0.01		100	
RO reject (SR2)	30% inlet flowrate of RO	0.005		200	
Wet I (SR3)	250	1		1	
Wet II (SR4)	200	2		0.5	
Litography (SR5)	350	3		0.3333	
CMP I (SR6)	300	0.1		10	
CMP II (SR7)	200	2		0.5	
Etc (SR8)	280	0.5		2	
Cleaning (SR9)	180	0.002		500	
Scrubber (SR10)	300	0.005		200	
Ultra pure water (UPW)	To be determined	18		0.0556	
Municipal fresh water (FW)	To be determined	0.02		50	

In addition, as discussed previously, 30% of the inlet flowrate for UF and RO systems is rejected as wastewater; thus, equations 13 and 14 are also included in the optimisation model.

$$F_{R}^{UF} = 0.3 \times F_{In}^{UF} \tag{13}$$

$$F_{R}^{RO} = 0.3 \times F_{In}^{RO} \tag{14}$$

Following the proposed automated targeting technique, the optimisation model is solved subject to minimise total municipal fresh water, F_{TFW} (fresh water for pre-treatment, F_{In}^{UF} and fresh water for water network, F_{FW}) with constraints of Equations 3 – 5. The PRCD for Example 2 is generated in Figure 7. Note that the total municipal fresh water and wastewater targets are found to be 3095 t/h (= 928.5 t/h + 649.95 t/h + 1516.55 t/h) and 2205 t/h, respectively. Note that all the municipal water is passed through the pre-treatment system to form UPW. The optimal water recovery scheme for Example 2 is showed in Figure 8.

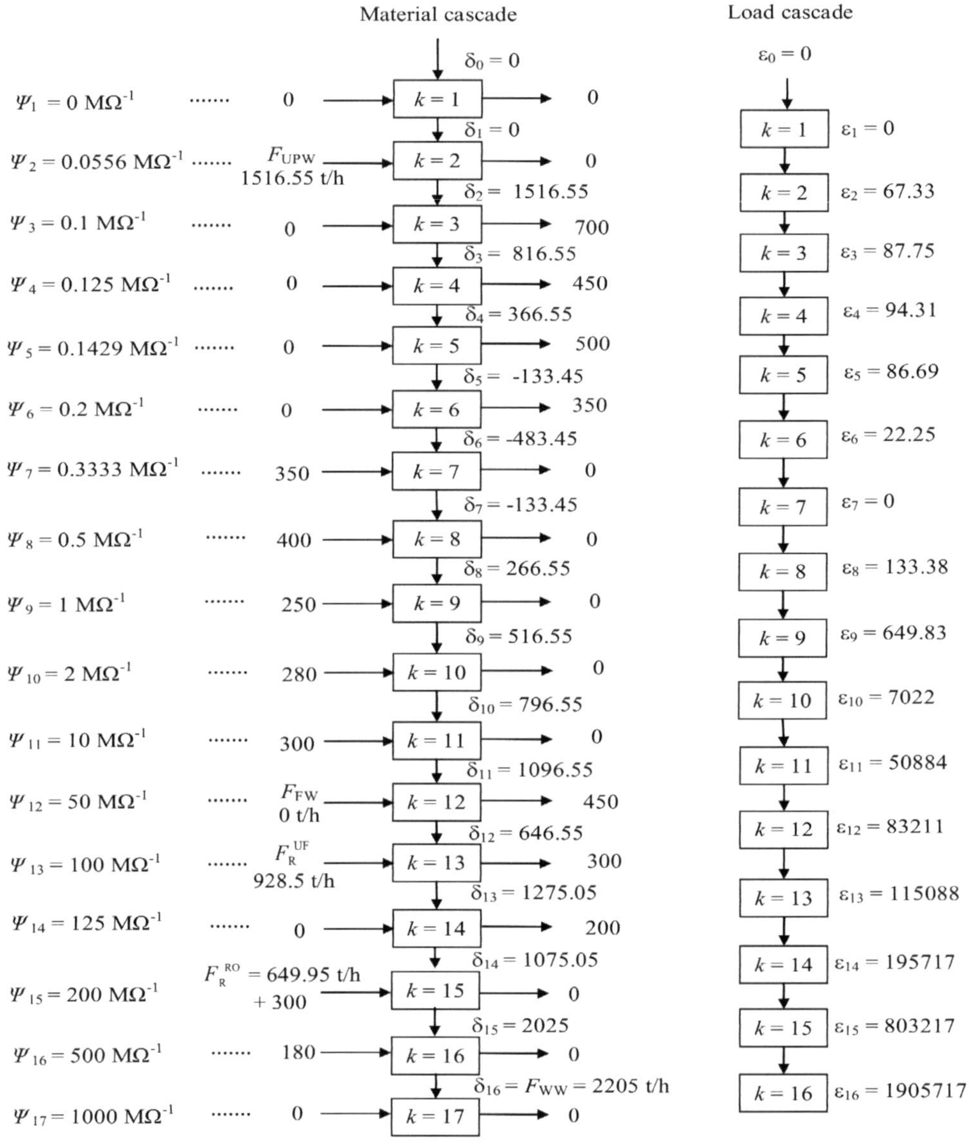

Figure 7 PRCD for Example 2

CONCLUSION

In conclusion, this work presents the extension of automated targeting technique to establish the resource targets within a property integration framework. The automated targeting combines the advantages of both insight-based and mathematical optimisation approaches. Two examples have been solved using this approach.

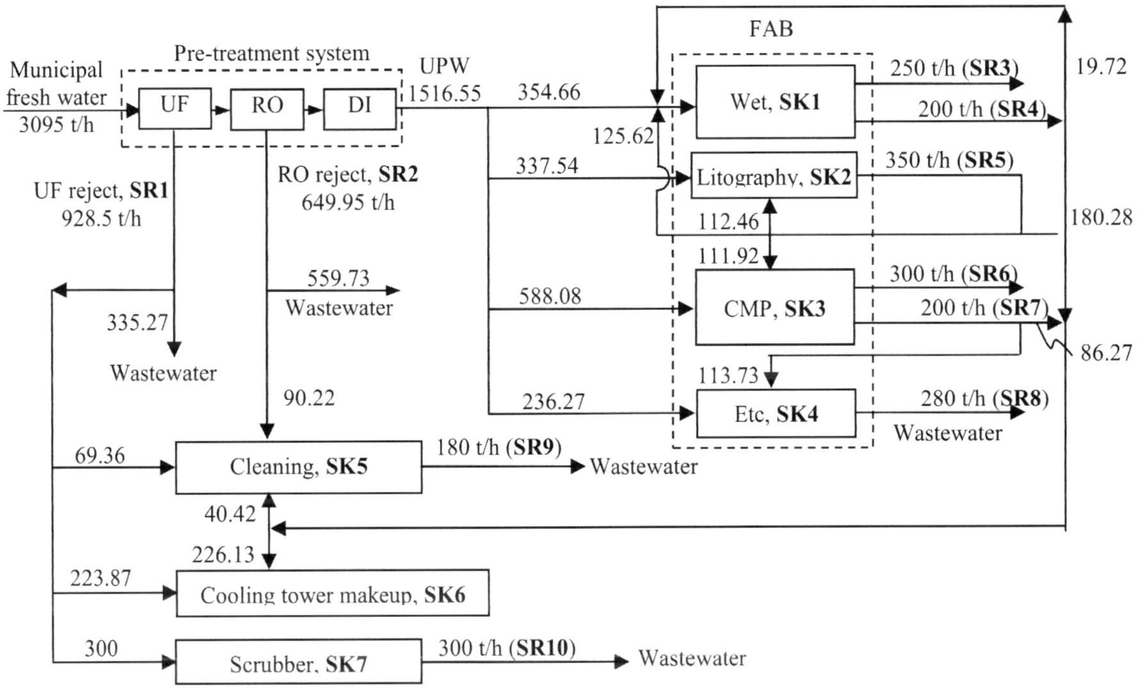

Figure 8 Optimal water recovery scheme for wafer fabrication process (Example 2).

REFERENCES

1. Almutlaq, A.M., Kazantzi, V., El-Halwagi, M.M., 2005. An algebraic approach to targeting waste discharge and impure fresh usage via material recycle/reuse networks. Clean Technology and Environmental Policy 7 (4), 294-305.
2. Bandyopadhyay, S., Source composite curve for waste resuction. Chem. Eng. J. 125 (2006) 99-110.
3. El-Halwagi, M.M., 1997. Pollution Prevention through Process Integration: Systematic Design Tools. Academic Press, San Diego.
4. El-Halwagi, M.M., 2006. Process Integration. Elsevier Inc., Amsterdam, Netherlands.
5. El-Halwagi, M.M., Springgs, H.D. 1998. Solve deign puzzles with mass integration. Chemical Engineering Progress 94, 25-44.
6. El-Halwagi, M.M., Gabriel, F., Harell, D., 2003. Rigorous graphical targeting for resource conservation via material recycle/reuse networks. Industrial and Engineering Chemistry Research 42, 4319-4328.
7. El-Halwagi, M.M., Glasgow, I.M., Eden, M.R., Qin, X., 2004. Property integration: componentless design techniques and visualization tools. A.I.Ch.E. Journal 50 (8), 1854-1869.
8. Foo, D.C.Y., Kazantzi, V., El-Halwagi, M.M., Manan, Z.A., 2006. Surplus diagram and cascade analysis technique for targeting property-based material reuse networks, Chemical Engineering Science 61(8), 2626-2642.
9. Kazantzi, V., El-Halwagi, M.M., 2005. Targeting material reuse via property integration. Chemical Engineering Progress 101 (8), 28-37.
10. Kuo, W.C.J., Smith, R., 1998. Design of water-using system involving regeneration. Transactions of the Institute of Chemical Engineers, Part B,vol 76, 94-114.

11. Manan, Z. A., Tan, Y. L., Foo D. C. Y., 2004. Targeting the minimum water flowrate using water cascade analysis technique. AIChE J. 50(12) 3169-3183.
12. Ng, D.K.S., Foo, D.C.Y., Tan, R.R., Tan, Y.L., 2007. Ultimate flowrate targeting with regeneration placement. Chemical Engineering Research and Design, Trans IChemE Part A, vol.85 (A8), 1-5.
13. Ng, D. K. S., Foo, D. C. Y., Tan, R. R., Automated targeting technique for resource conservation networks. An international Conference on Water & Wastewater (Asia Water 2008) (2008).
14. Pau, C. H., 2007. Property integration network with regeneration placement. Thesis of final year research project. Curtin University of Technology.
15. Shelley, M.D., El-Halwagi, M.M., 2000. Componentless design of recovery and allocation systems: a functionality-based clustering approach. Computers and Chemical Engineering 24, 2081-2091.

521a Introduction to Chemometrics -- Design of Experiments and Principal Component Analysis

Dongsheng Bu and Andrew Chu. CAMO Software Inc., One Woodbridge Center, Suite 319, Woodbridge, NJ 07095

Design of Experiments (DOE) is an efficient tool to develop process understanding and to construct a design space. DOE varies the values of studied factors in parallel and the experiments are selected to adequately cover the entire range of operations. The DOE protocol demands statistical validity, so costly development can be restricted to significant results with great confidence. Pharmaceutical manufacturing typically involves a series of process steps, each with many variables that can have a major impact on critical responses. Principal Component Analysis (PCA) is to decompose multivariate data with correlated measurements into a new set of uncorrelated principal components. The PCA results are often presented with graphs plotting the projections of the observations onto the components as scores, and the projections of the measurement variables as loadings. The importance of each component is expressed by the proportion of the variance explained.

This paresentation utilized a pharmaceutical active ingredient (API) synthesis as an example to develop process understanding by DOE and PCA. The purpose of this study was to identify critical process variables and detect possible interactions between variables such as temperature, reagent concentration and solvent type, which would affect yield, purity, isomer ratio, and Raman measurement quality. Main effects and interactions were assessed by ANOVA, followed by Response Surface modeling to study the relationship between process variables and property responses. Principal component analysis was used to analyze the variability in the process and isomer product physical properties.

521b Data Exploration Analysis In Process Development: Application to API Characterization

Jose E. Tabora, Jessica Defreese, Prashant Deshpande, and Jean Tom. Bristol-Myers Squibb, One Squibb Drive, New Brunswick, NJ 08901

The characterization of the physical properties of the API is an enduring challenge in pharmaceutical process development. Multiple attributes such as particle size distribution, specific surface area, bulk and tapped density are routinely measured in efforts to define the acceptability criteria for subsequent processing. In addition, due to the complexity of the processes involved, a phenomenological relationship between the process parameters from crystallization and drying and the resulting physical attributes of the isolated material is generally intractable. In this work we present the application of data exploration analysis and chemometrics to facilitate the characterization of the API powder properties in relationship to process parameters. Material isolated from multiple lots of Pilot plant scale processing was characterized by up to 12 measured properties. Multivariate analysis (PCA and multiway-PCA) was applied to the descriptors and to the batch processing profiles to aide visualization of a large data set. The trends identified from data mining were evaluated in the context of known physical processes.

521c Scale-up a Complex Pharmaceutical Batch Process Using Mid-IR: The Role of Simple Peak Height Profiling, Partial Least Squares (PLS) Modeling and Dynamic Principal Component Analysis (PCA)

Shan Lin, James Marek, and Samrat Mukherjee. GPRD Process Engineering, Abbott Laboratories, 1401 Sheridan Rd, North Chicago, IL 60064-6292

Monitoring batch process is essential for pharmaceutical production quality and reducing cycle time. In this study, batch mid-infrared (MIR) spectra based partial least square (PLS) modeling, principal component analysis (PCA) and simple peak height profiling were used for monitoring, prediction and characterizing batch reactions. We studied many batches of Triflation reactions including six production batches and acquired hundreds of MIR spectra directly from each batch. Performing PLS modeling had been able to accurately predict batch reaction concentrations and the reaction end point with the prediction sensitivity as low as < 1 PA% of starting R-OH residues. Performing PCA could reduce hundreds of multi-dimensional spectra data matrix into a few dimensional score plots for visualizing batch concentrations, end point and other batch variations with limited calibration efforts. Temperature involved single peak and multiple-peak heights regression profiling could be a quick approach for batch quantitative measurement and end point detection. This paper also described a quick method for transferring lab calibration data into pilot plant scale with good prediction accuracy.

Key word: Batch process, MIR, PLS, PCA, regression Profiling

521d Application of Chemometrics in Process Monitoring of Synthesizing Preferred Isomer

Dongsheng Bu and Andrew Chu. CAMO Software Inc., One Woodbridge Center, Suite 319, Woodbridge, NJ 07095

A pharmaceutical manufacturing process to produce a preferred isomer was studied using Raman spectroscopy. One of the objectives is to achieve better control throughout the process, as a way for improving product quality. In order to accurately characterize the preferred isomer and undesired isomer during synthesis process, appropriate Raman data pre-treatments are necessary to correct baseline drifts and pathlength variations. Multivariate Curve Resolution (MCR) is applied to monitor relative concentrations of a major reagent and the isomers over time. Principal Component Analysis (PCA) score plots of different batches are useful in the following changes of measured spectra and to study batch-to-batch difference.

521e Utilization of Single Crystal X-Ray Studies of Process Relevant Pharmaceutical Crystal Forms to Support Qbd

John D. DiMarco, Michael A. Galella, Raymond P. Scaringe, Carolyn J. Pommier, and Jack Z. Gougoutas. Solid State Chemistry, Analytical R&D, Bristol-Myers Squibb, Route 206 & Province Line Road, Princeton, NJ 08543

Ideally, the active pharmaceutical ingredient has a single well-characterized and relatively stable crystal structure. However, complications can arise for molecules that crystallize in several solvated and/or neat forms which may de-solvate/transform and/or interconvert during manufacturing processes. Single crystal studies of these forms define their structure/composition and moreover allow PXRD simulations and correlation of spectra with structure (Raman, IR) which can be used in the determination of thermodynamic relationships, the mapping of design space, and ultimately to improve process control. Some recent examples will be presented, with emphasis on process relevant forms of dasatinib.

595a Selecting a Capstone Product Design Project

Michael Hill, M. Hill & Associates, PO Box 622, Wyckoff, NJ 07481, Sanat K. Kumar, Department of Chemical Engineering, Columbia University, 500 W 120th St, Mudd 814, New York, NY 10027, and **Lynn H. Lander**, Lander & Associates, LLC, PO Box 201, Harrington Park, NJ 07640-0201.

Columbia University has several years of experience with its undergraduate capstone product design course, requiring students to complete a semester-long product design project. Among our previously assigned projects were the design of biodegradable roadway deicers, drinking water purification devices, portable hemodialysis devices, and second generation biofuels. Looking back at previous projects, we have been able to extract important lessons about project selection and assignment. This paper will discuss these lessons, including the various factors that make for a successful project.

595b Product Design Using Reverse Engineering

Maria Nydia Ruiz Felix and Robert P. Hesketh. Chemical Engineering, Rowan University, 201 Milluca Hill Road, Glassboro, NJ 08028

Considerable interest has focused on the design of new products that identify processes, ingredients and systems that are environmental friendly. Chemical products can be manufactured by following different routes. Every route has a planning, research and development cycle. The reverse engineering is a tool used in the education field to understand the function and construction of a product or machine. This method will be used to design a thermochromic label. Because of their simple design and low cost, irreversible thermochromic sensors can be used in many applications in the food, textile, and commercial product industries. Initially thermochromic sensors appear white, when the specific temperature is reached, the sensor color changes. This indicates that the substance the sensor was placed on reached the temperature of activation.

Commercial sensors were tested using reverse engineering to make the Rowan sensors. Parameters such as activation temperatures, activation time, amount of thermochromic mixture in the sensor, mixture quality, and particle size were taken into consideration. The labels were produced and characterized in a temperature range of 61°C to 97°C and 106°C to 131°C. The performance characteristics of Rowan's sensors are similar to the ones currently on the market.

595c Experiences with Typical Product Design Projects

Warren D. Seider[1], John C. Crocker[2], Talid Sinno[1], and Scott L. Diamond[1]. (1) Chemical and Biomolecular Engineering, University of Pennsylvania, Philadelphia, PA 19104, (2) Department of Chemical and Biomolecular Engineering, University of Pennsylvania, 220 South 33rd Street, Philadelphia, PA 19104

Over the past several years, we have formulated product-design projects for groups of three and four undergraduate students. These projects, which are executed during the second semester of the senior year, include designs of labs-on-a-chip for the high-throughput screening of pharmaceuticals, the CVD of electronic materials, and batch processes to manufacture pharmaceuticals. In these projects, emphasis shifts to the 3D configuration of the product, the molecular simulations of surface roughness using kinetic Monte-Carlo methods, and/or the scheduling of small batch processes - away from the simulation of commodity chemical processes.

595d Design of Pharmaceutical Products: Experiences at Rose-Hulman

David C. Miller, Rose-Hulman Institute of Technology, Chemical Engineering Department, 5500 Wabash Ave, Terre Haute, IN 47803

For the past seven years, the design of pharmaceutical products has been a popular capstone project among senior chemical engineers at Rose-Hulman Institute of Technology. In addition to product design, these projects also contain elements of process design. This paper will examine how the product design elements of these projects enhance the overall learning experience for the students.

Product Design with the Aid of Finite Element Programs In Capstone Design Projects

Aydin Sunol, Department of Chemical Engineering,
University of South Florida, 4202 East Fowler Avenue, ENB 118,
Tampa, FL 33620

Abstract

The capstone chemical engineering design course projects are based on products as opposed to process. Furthermore, we are seeing increased reliance on computationally more advanced techniques such as finite element methods integrated reliably in commercially available software. In this presentation, we want to share our experience in use of COMSOL throughout several product design projects.

Introduction

The chemical engineering design effort that historically started with a focus on unit processes, equipment and plants soon transformed to concentrate around the processes and systematically evolved through computational tools such as flowsheet simulators as well as process synthesis methodologies. The concern for environment and safety became part of capstone courses in the past two decades. With evolving and expanding role of our profession in areas where value added differentiated products dominate and increased proximity of curricula to life sciences, the incorporation of product design in our curriculum was inevitable. This is in part brings forth need for incorporation of the supply chain and entire life cycle of the products in the design. The integration of process and product design would allow smooth flow of information from the design stage through the manufacturing stage. The integration of process and product design coupled with availability of software that can be effectively used for both necessitates utilization of computation tools such as finite element imbedded into mathematical software.

Capstone Design Course Series

Chemical Engineering programs usually have either one or two course design series. Probably like many institutions, at the University of South Florida, over the last three decades, we alternated between one course and two course series. At present, we have a first course where we focus on foundations of 'Product and Process Engineering' with a well defined design project. The second course 'Product and Process Design' is primarily centered on major design projects.

The first course was originally developed to teach strategies of process engineering in a systematic fashion where concepts were integrated through a project. Formal coverage of process economics, optimization, and uncertainty & risk analysis principles was coupled with a well defined simple design project with two to three degrees of freedom. The three credit course implied two hours per week of lecture time and three hours per week of tutorials and problem sessions.

The transformation to incorporate product design was gradual. Early to late nineties, we incorporated products primarily through market analysis. The economic analysis coverage was modified to incorporate key microeconomic concepts such as utility theory, supply demand analysis and pricing using game theory. In the past several years, the projects had a product design components and process design projects were modified accordingly. For example, a natural product extraction process can easily be modified to include multiple components and/or encapsulation stage. The other implications were choosing examples to reflect both the process and product realities. In that context, linear programming coverage touched on supply chain optimization. We also incorporated batch scheduling to deal with multi-product batch plants.

The second course, capstone design course, is centered on the projects that have varying degree of process or product emphasis but usually students end-up integrating both. There still two lecture and one three hour problem sessions per week. Some of lecturing is on general review of principles in the context of design and is with the aid of a sample pre-worked design project. Others are on new topics such as safety, green engineering, process synthesis and more rigorous sizing methods. The problem sessions enable applications particularly with the aid of computational tools such as flowsheet simulators. The first third of the course is lecture and tutorial intensive. At the beginning of the semester, students form groups and choose projects that are either from a list which is in part industry initiated or AIChE contest problem or self designed. Groups of three to four students work together and are mentored by teaching assistants, faculty. or industrial consultant/ There are weekly meetings and project requirements include a written report, oral presentation and poster presentation/ The projects have an individual as well as group component. The projects used over the last few years include:

- Self heating food, cappuccino, wash towel, stadium seats
- Self cooling beverage
- Food grade phosphoric acid
- Encapsulated Nutraceuticals
- Green Sail Boat

- Energy efficient amphibian for Tampa Bay Area
- Water purification system for Traveler
- Green Refrigerant
- LED plant
- Alcohol Free Beverage of your choice
- Targeted delivery drug of your choice
- Time release drug of your choice
- Wood-polymer composites
- Fuel cell based applications
- Dialysis Machine
- Chiral Separation Plant

The entire group is responsible from the design and following group report that includes the design and feasibility analysis. The group reports followed up with individual components of design and starts after the group report is submitted for evaluation. The individual report is based on detailed design of an element of the group report. The group has an oral presentation that is open to public. The last presentation is a poster presentation that is judged by the advisory board faculty and graduate students.

Finite Element Programs

The finite element programs benefit not only the product design element of the course but also process/equipment design. Historically, computation fluid dynamic packages were part of equipment design before product design. We found that distributed models provide a natural link between the engineering science courses such as transport phenomena and the design sequence.

Our students are introduced and solve a common tutorial problem in the workshop component of the course. They also use it in their product design as well as individual equipment design which may be considered product as well. Our students use COMSOL that interfaces well with MATLAB.

In the product design projects, we found them COMSOL to be particularly useful with timed release, targeted delivery, and self heating/cooling product design projects. This was in part due to the fact that models were simpler and readily available. Furthermore, these models were essential for prototyping the product. .

632a Product Driven Process Design Methodology

Peter M.M. Bongers, Chemical Engineering and Chemistry, Eindhoven University of Technology, PO box 513, Eindhoven, 5600 MB, Netherlands and Cristhian P. Almeida-Rivera, Process Science, Unilever R&D Vlaardingen, Olivier van Noortlaan 120, POBox 114, Vlaardingen, 3130AC, Netherlands.

In the last ten years much more processes are being reported to be desinged through a process synthesis approach. It has been recognised during those years that (i) processes for structured products are much more difficult to design through process synthesis; (ii) processs synthesis is disconnected from product development. In a response to these shortfalls a number of authors have described that these gaps need to be filled, however no methodology extension has been proposed.

In this work, we will present extensions to the conceptual process synthesis methodology to include the (structured) product design. The whole design methodology spans from how the new product can enlight the consumer, financial and supply chain boundary conditions, through to an optimal flowsheet able to produce the desired product cost effectively.

References

Douglas, J., 1988, Conceptual Design of Chemical Process, McGraw-Hill, New York.

Meeuse, F. M., Grievink, J., Verheijen, P. J. T., and Stappen, M. L. W. W., 1999, Conceptual design of processes for structured products, in Fifth conference on Foundations of Computer Aided Process Design (AIChE, Breckenridge, USA).

Hill, M., 2004, Product and Process Design for Structured Products, AIChE Journal, 50(8): 1656-1661.

Gone Fishing !

A CRACKER STORY

GARY ISRAEL
PROCESS SOLUTIONS OF NY LLC
SEPTEMBER 19, 2008

1995

7 CRACKER PRODUCTION LINES – 3 PLANTS

20 + SHIFTS /LINE / WEEK

ROOM FOR ONE ADD'TL LINE – $18MM

MAX OUT - PRE PACKAGING 

COMPANIES LARGEST SALES SKU'S

'INTERNAL ONLY' PROTECTION – NO CO-MNFG

EXPERTS - " ***LINE SPEED MAXED OUT***! "

BUSINESS &

TECHNICAL CHALLENGE

DOUBLE BUSINESS BY 2000

GO BACK TO FISH CRACKER SCHOOL

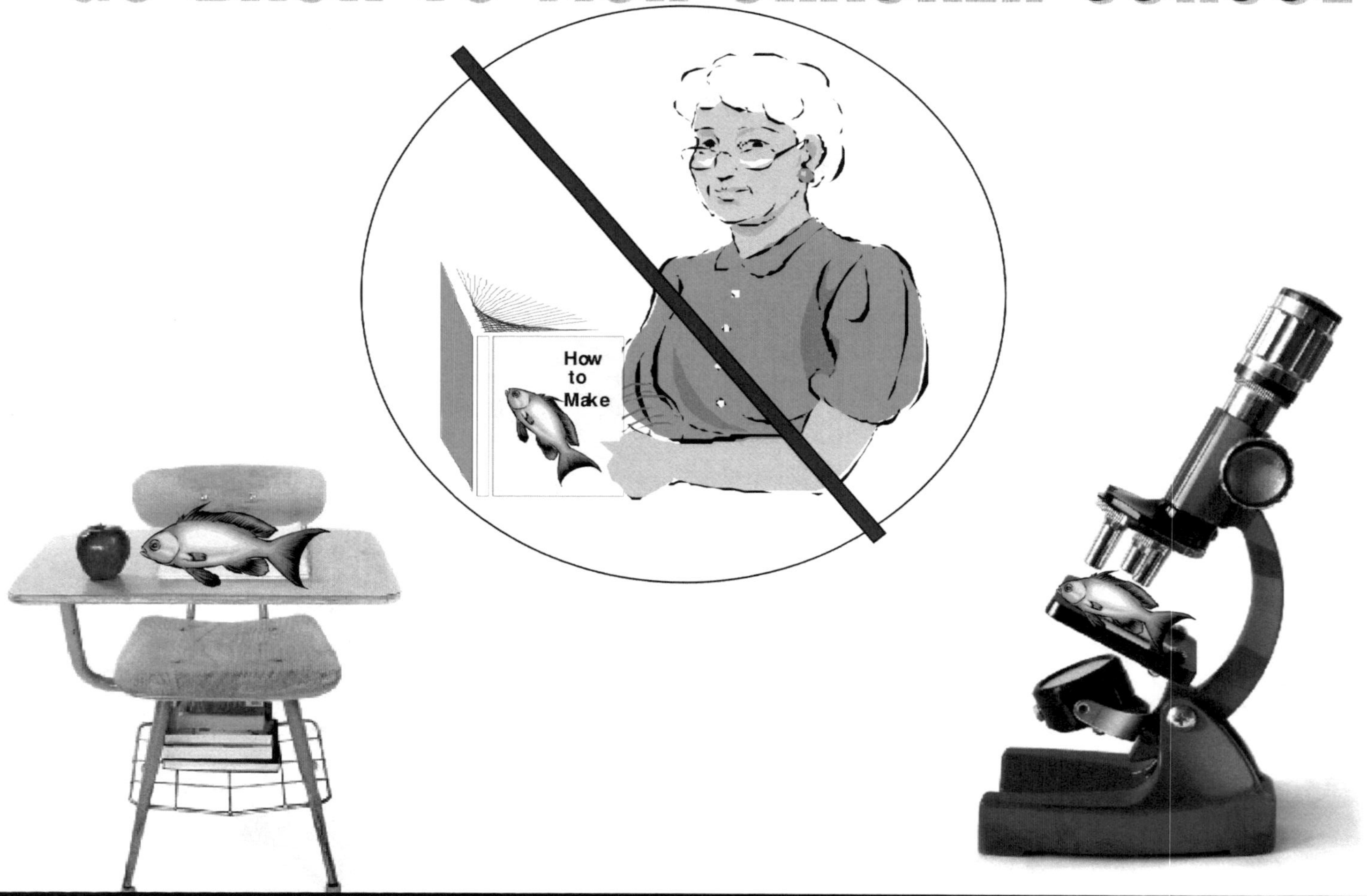

GARY ISRAEL
PROCESS SOLUTIONS OF NY LLC
SEPTEMBER 19, 2008

SNACK CRACKER PRODUCTION

Mixing & Fermentation

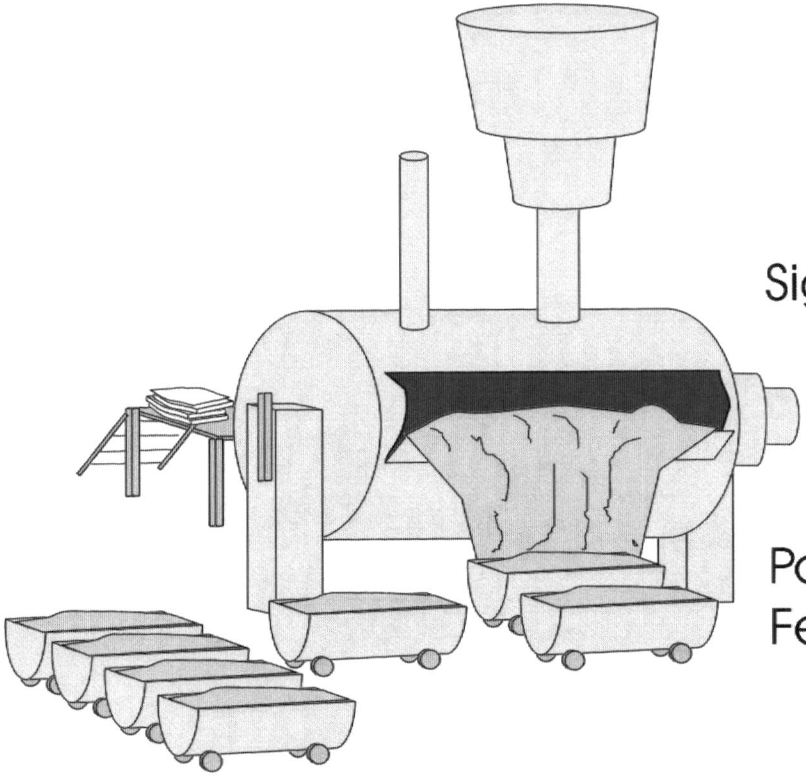

Sigma Blade Mixer

Portable Dough Trough
Fermentation

$\frac{\text{Flour}}{\text{Starch \& Sugar}} + \frac{\text{Yeast}}{}$ ⟶ Ethanol
Sugar Alcohols
CO_2

A CRACKER STORY

GARY ISRAEL
PROCESS SOLUTIONS OF NY LLC
SEPTEMBER 19, 2008

Mixing

Stage 1: Incorporation
Stage 2: Uniformity
Stage 3: Development

Flour
Water
Cheese
Shortening
Emulsifiers
Yeast
Sugars

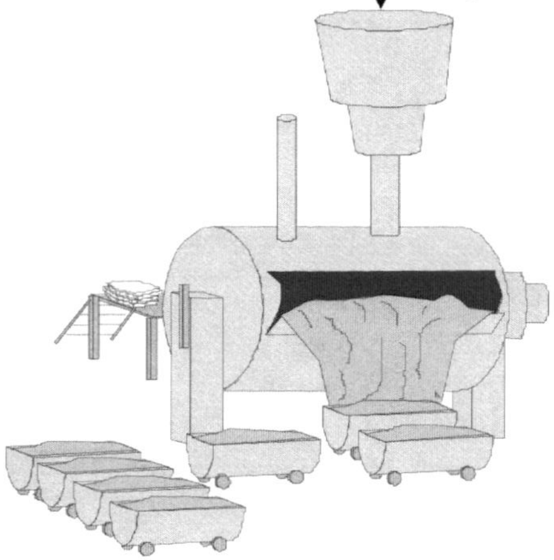

Mixing - Development

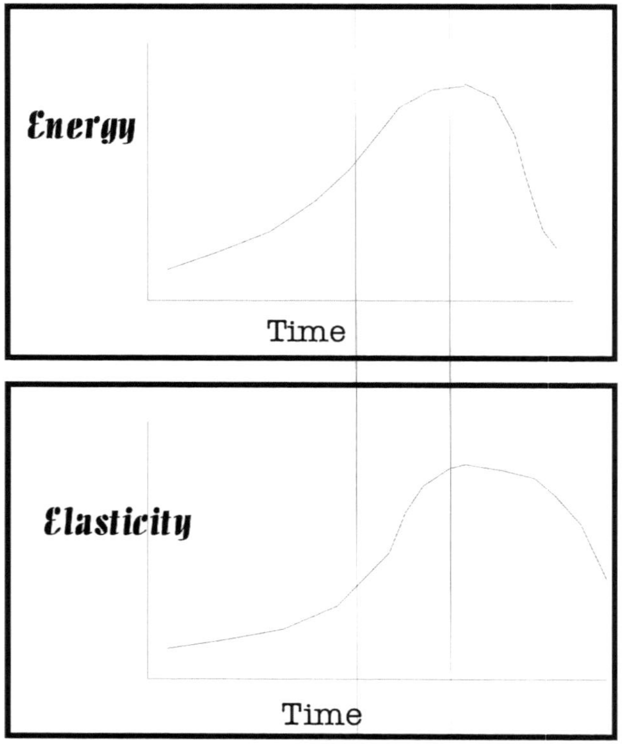

Crackers

A CRACKER STORY

GARY ISRAEL
PROCESS SOLUTIONS OF NY LLC
SEPTEMBER 19, 2008

SNACK CRACKER PRODUCTION

MAKEUP PRIOR TO BAKING

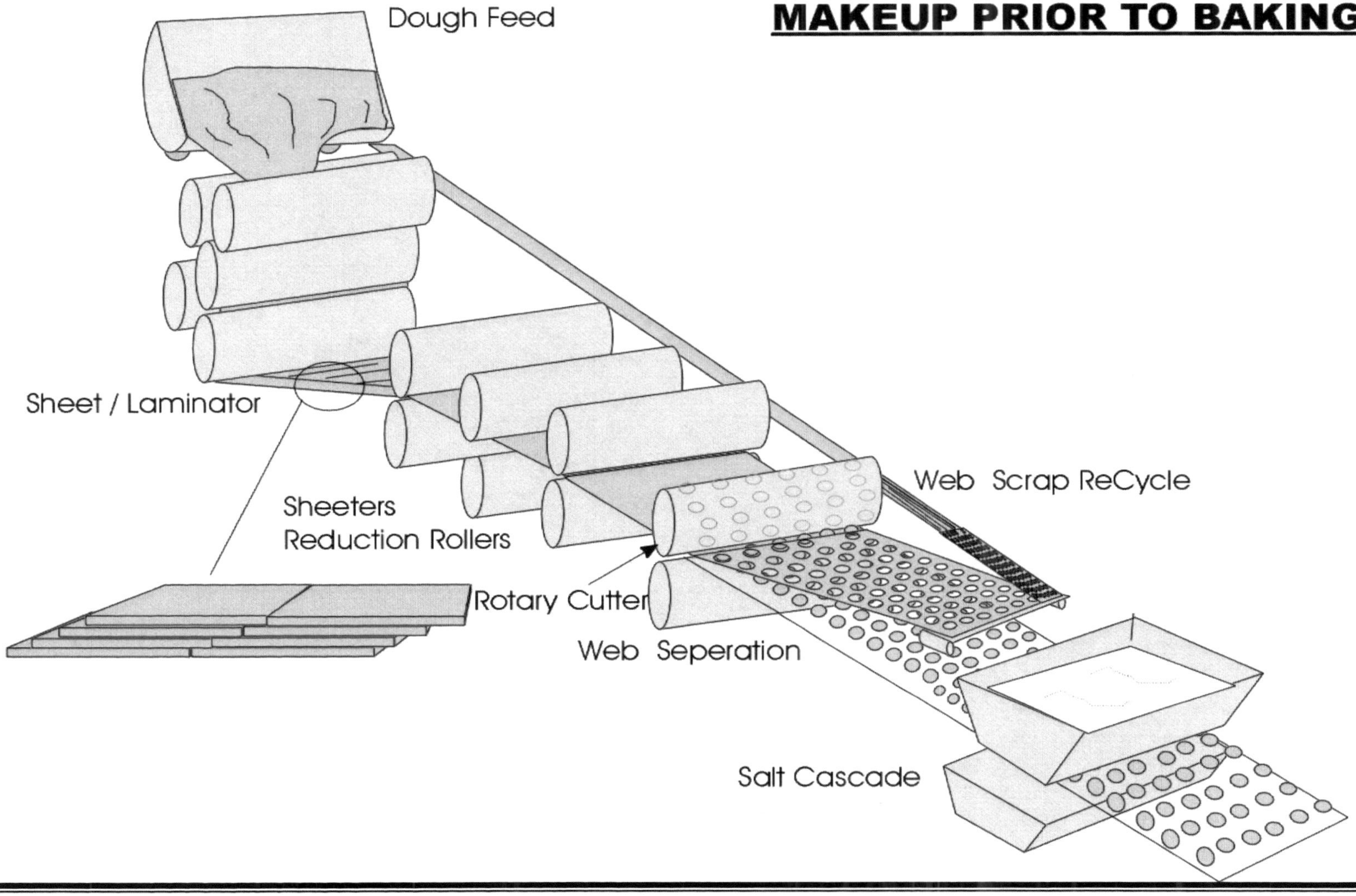

GARY ISRAEL
PROCESS SOLUTIONS OF NY LLC
SEPTEMBER 19, 2008

Snack Cracker Production

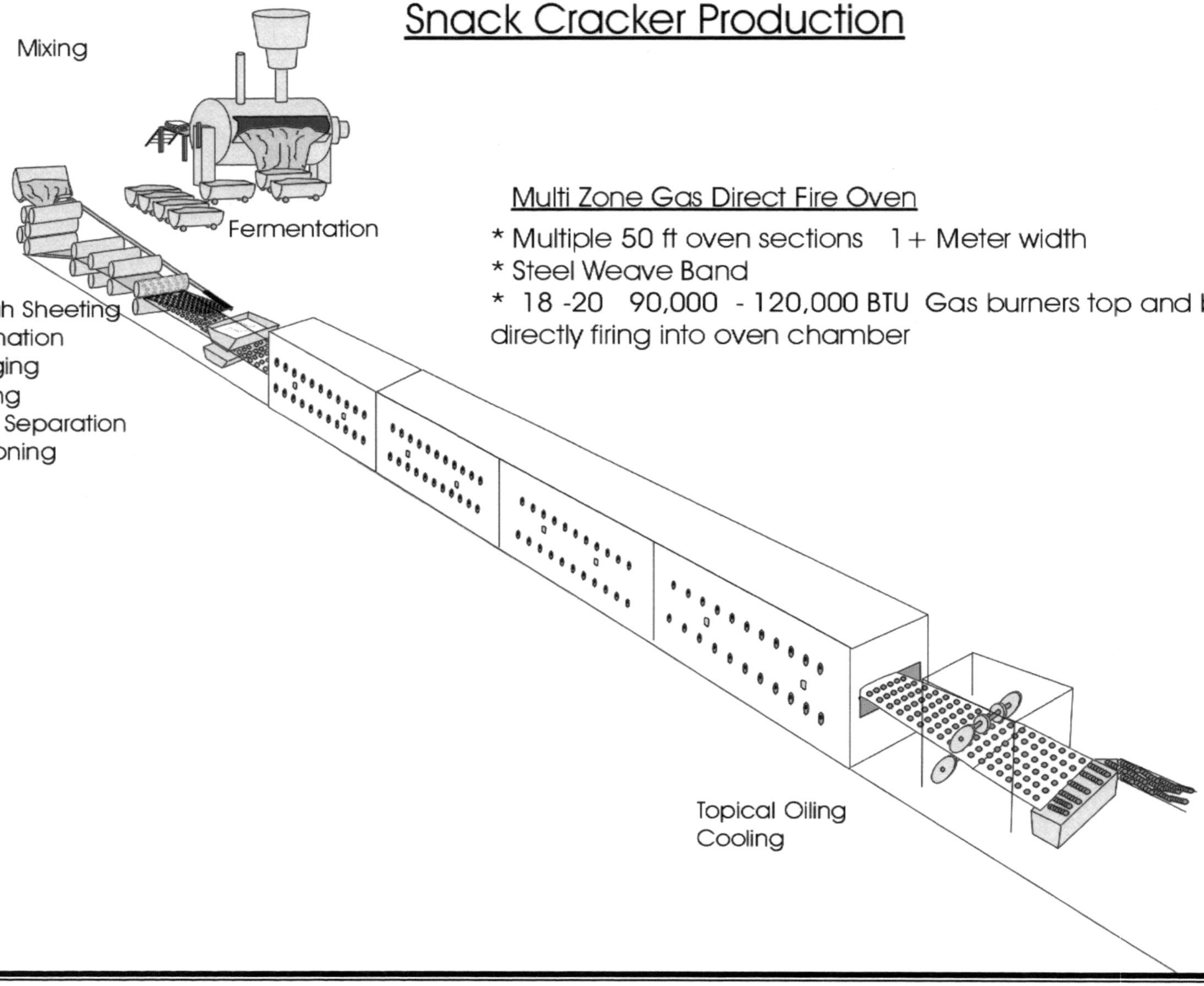

Multi Zone Gas Direct Fire Oven

* Multiple 50 ft oven sections 1 + Meter width
* Steel Weave Band
* 18 -20 90,000 - 120,000 BTU Gas burners top and bottom directly firing into oven chamber

GARY ISRAEL
PROCESS SOLUTIONS OF NY LLC
SEPTEMBER 19, 2008

CONVENTIONAL CRACKER BAKING

GDF: GAS DIRECT FIRE

Ribbon Burners in Product Chamber - Individual Control

Dense Steel Weave Oven Band

Multiple Baking Zones

Humidity Control Exhaust Stacks (fan & damper)

GARY ISRAEL
PROCESS SOLUTIONS OF NY LLC
SEPTEMBER 19, 2008

Snack Cracker Production

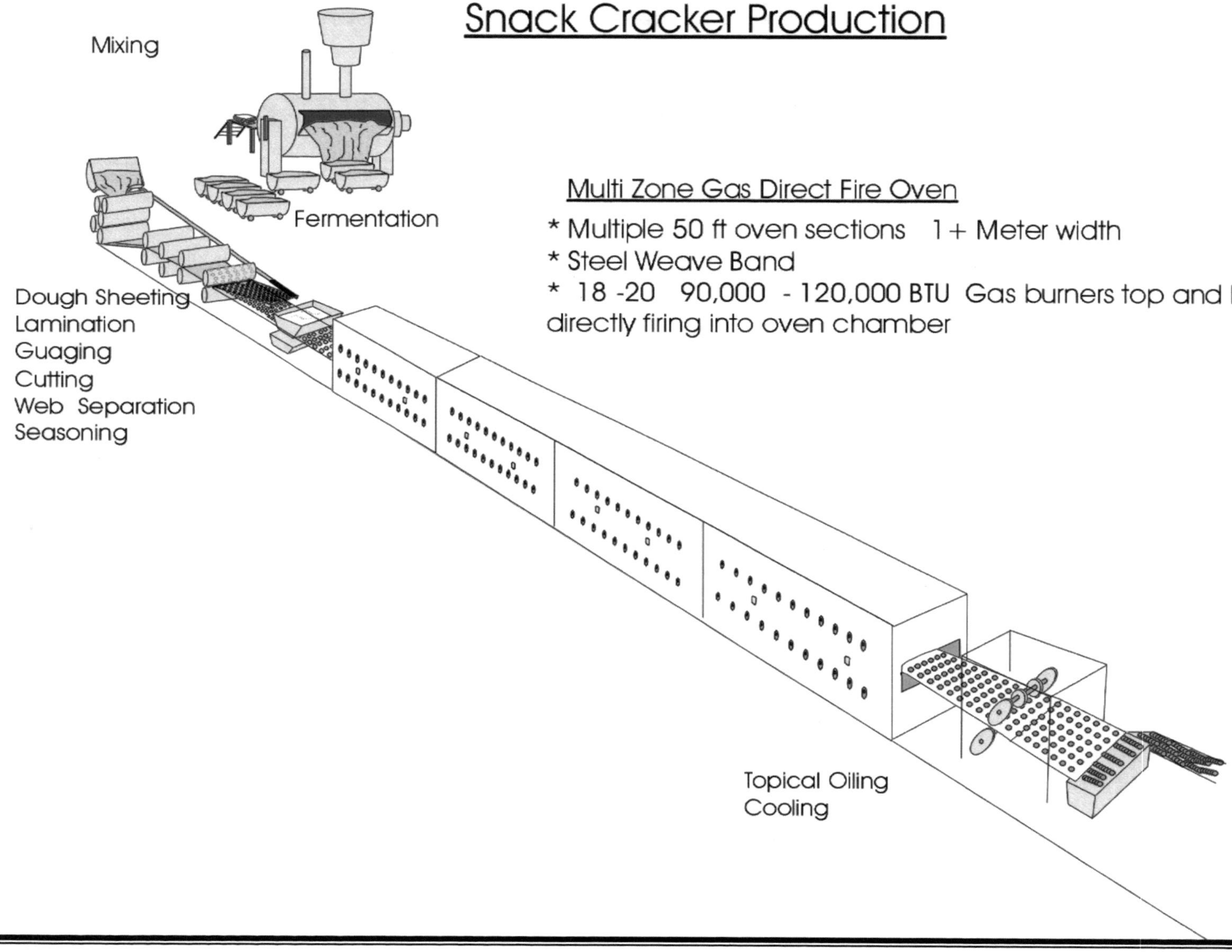

Multi Zone Gas Direct Fire Oven

* Multiple 50 ft oven sections 1 + Meter width
* Steel Weave Band
* 18 -20 90,000 - 120,000 BTU Gas burners top and bottom directly firing into oven chamber

A CRACKER STORY

GARY ISRAEL
PROCESS SOLUTIONS OF NY LLC
SEPTEMBER 19, 2008

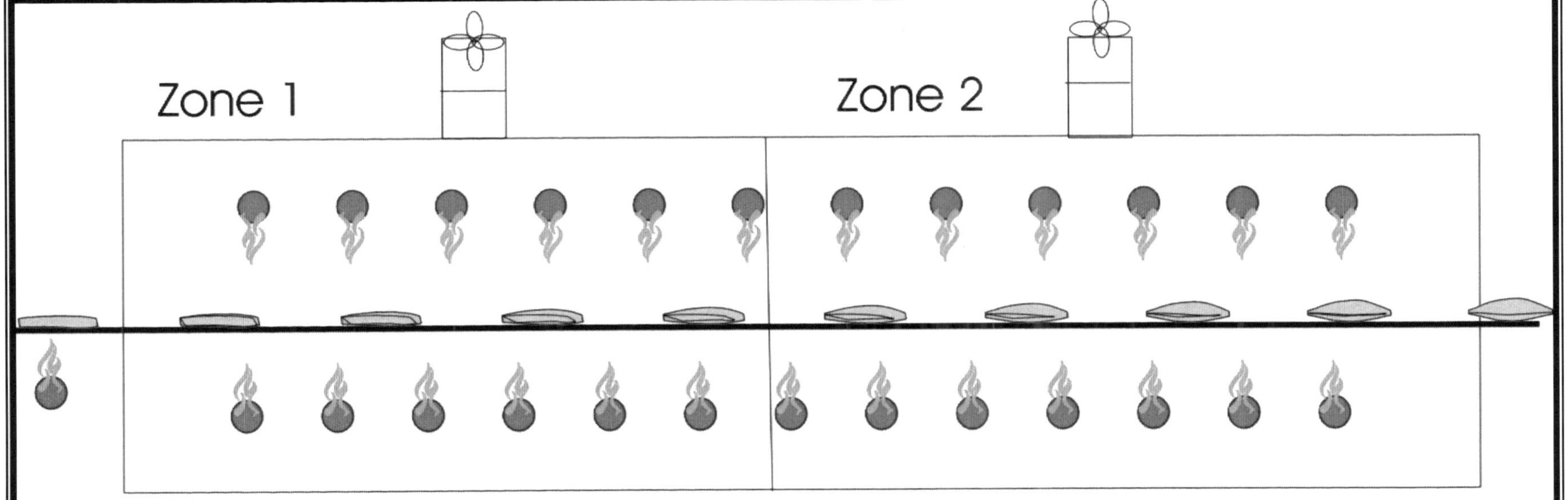

Product

* Internal Temperature and heat Increase
* Partial Pressure of Water in dough increases
* Evaporation at Top Surface
* Defusion Reduction at Surface (Skinning)

Oven

* Band Heating
* Humidity Increase - Closed Exhust Damper

Product

* Internal Temperature and heat Increase
* Defusion Reduction at Surface (Skinning) + Trapped steam starts layer separation
* Evaporation continues at Top Surface Product bows off band, Evaporation from bottom and sides. Puff begins at Z2.

Oven

* Band Heating - Band Temp increases
* Humidity Increase - Closed Exhust Damper

GARY ISRAEL
PROCESS SOLUTIONS OF NY LLC
SEPTEMBER 19, 2008

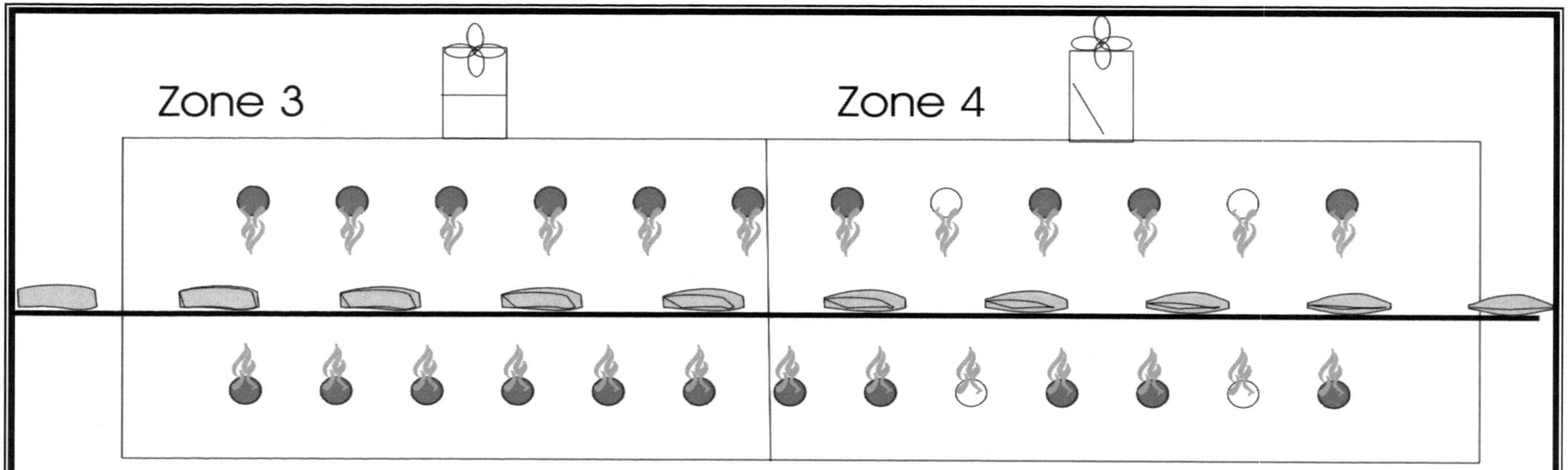

Product

* Primary purpose - Moisture redux (17% to 12%)

* Important to assure that product does not fully skin, and that scorching does not occur.

* Puff levels at maximum by end of Zone

Oven

* Band Heating
Heat carried into top of Zone 3 from Band heat in Zone #2.
* Top and Bottom Heat influx at Maximum

* Humidity Increase - Closed Exhaust Damper

Product

** Primary purpose - Moisture redux (127% to 6%)

* Product Structure set by mid Zone.

* Evaporation continues at Top Surface
Product bows off band, Evaporation from bottom and sides.

Oven

* Band Heating - Band Temp increased 75 deg

* Humidity decrease - Open Exhaust Damper Pull heat from Z 3, and reduce heat flow into Z5

GARY ISRAEL
PROCESS SOLUTIONS OF NY LLC
SEPTEMBER 19, 2008

Zone 5

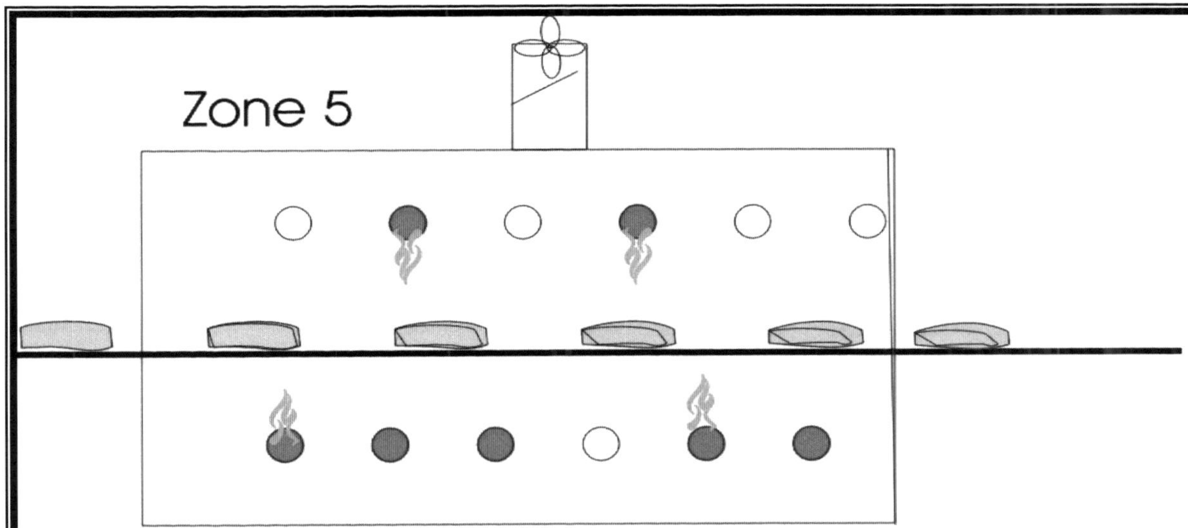

Product

* Primary purpose - Final Moisture redux (6% to 3%)

* Final Toast Flavor and color

* Important to assure that product does not scorch..

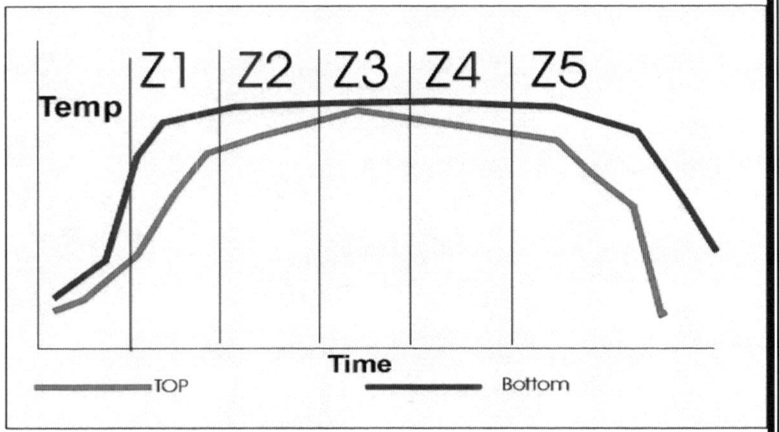

Oven

* Band Heating
  Heat carried into top of Zone 3 from Band
  heat in Zone #4.

* Humidity Decrease - Open Exhaust Damper

GARY ISRAEL
PROCESS SOLUTIONS OF NY LLC
SEPTEMBER 19, 2008

Snack Cracker Production

Step #1: Cutter Ring Optimizations

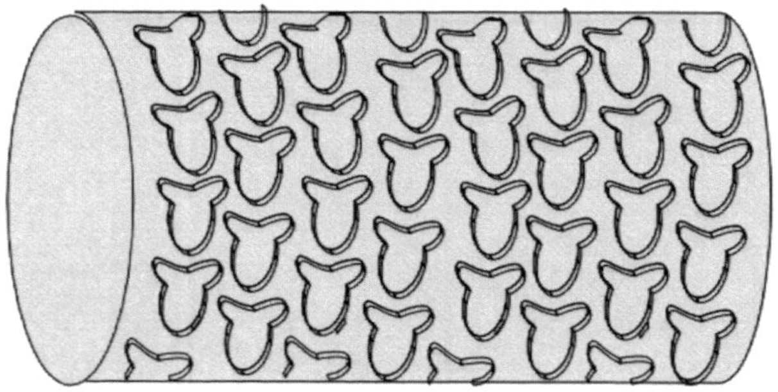

		#/Shaft	#/200' Oven @ 30RPM Crackers / Min	
Project Start	Plastic Cutters	490	14,700	
Stage 1	Brass Cutters	670	20, 100	+37%
Stage 2	Brass Cutters	873	26,190	+78%
Stage 3	Brass Cutters	912	27,360	+86%

Same 4.2 min Bake

GARY ISRAEL
PROCESS SOLUTIONS OF NY LLC
SEPTEMBER 19, 2008

Snack Cracker Production

Step #2: Bake Optimizations

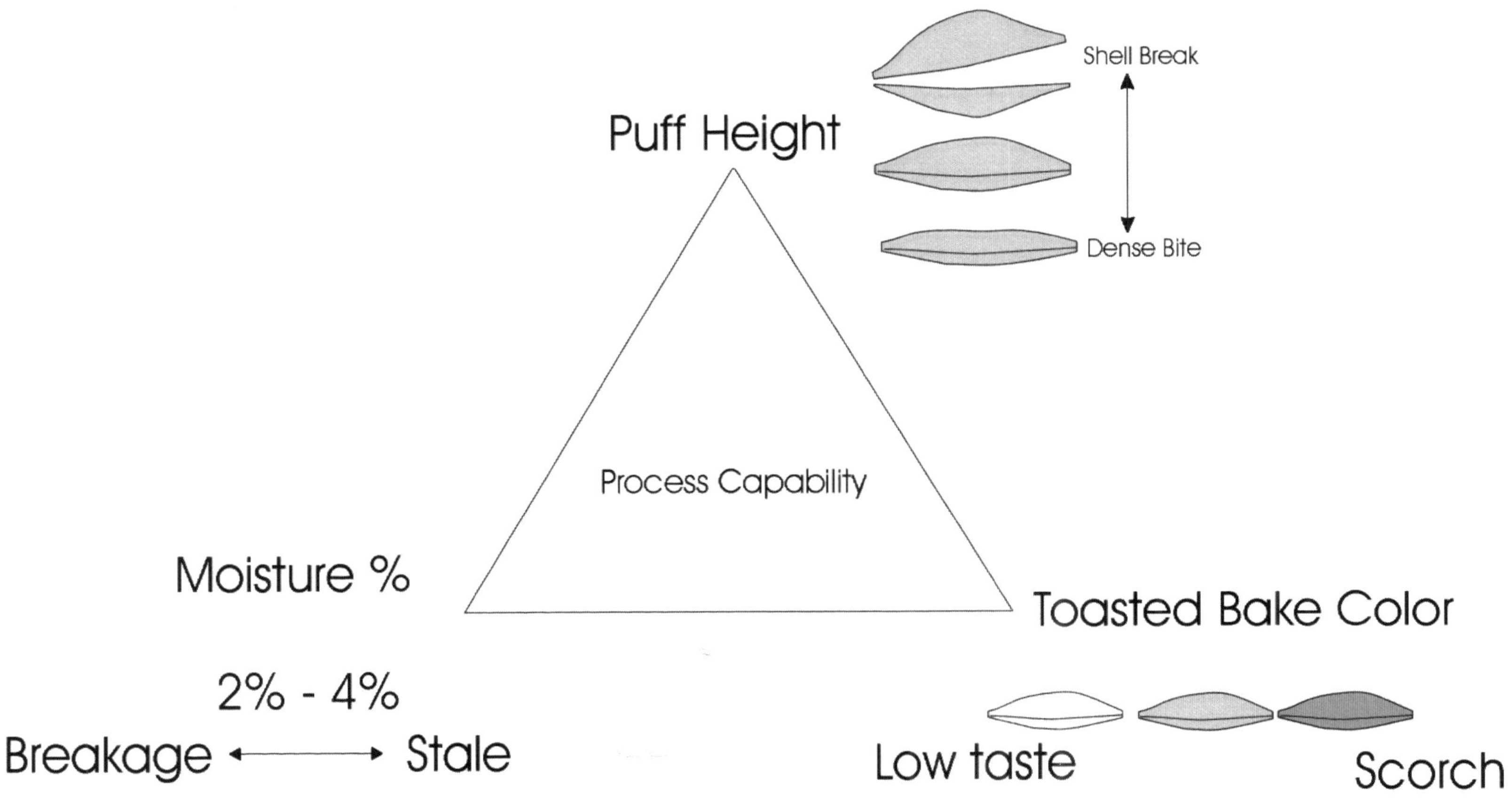

A CRACKER STORY

GARY ISRAEL
PROCESS SOLUTIONS OF NY LLC
SEPTEMBER 19, 2008

CHALLENGES TO DOUBLE

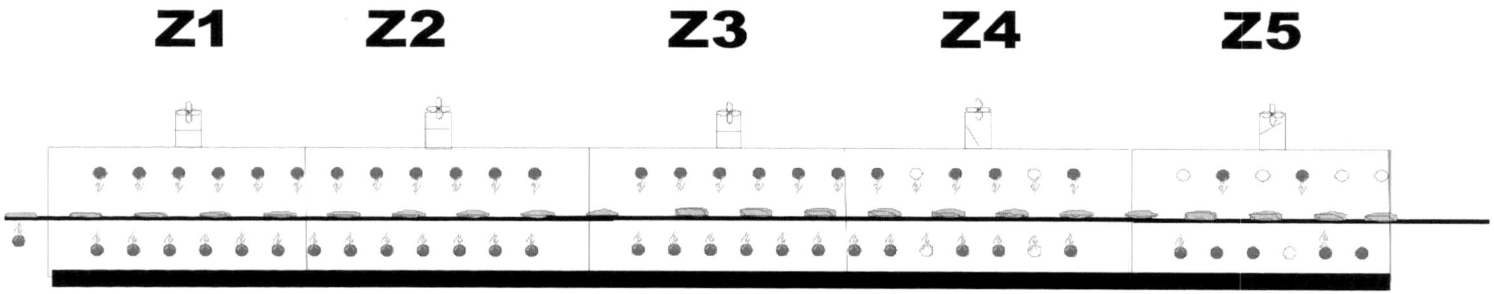

ORIGINAL 5 MIN. BAKE
- FULL PUFF ON 2.2 MIN
- MOISTURE EXIT

Z3	12%
Z4	6%
Z5	3%

ORIGINAL TARGET 2.5 MIN. BAKE
- FULL PUFF ON 1.2 MIN
- MOISTURE EXIT

Z3	12%
Z4	6%
Z5	3%

2 x HEAT TRANSFER

ALL HEATING VIA THERMAL GRADIENTS

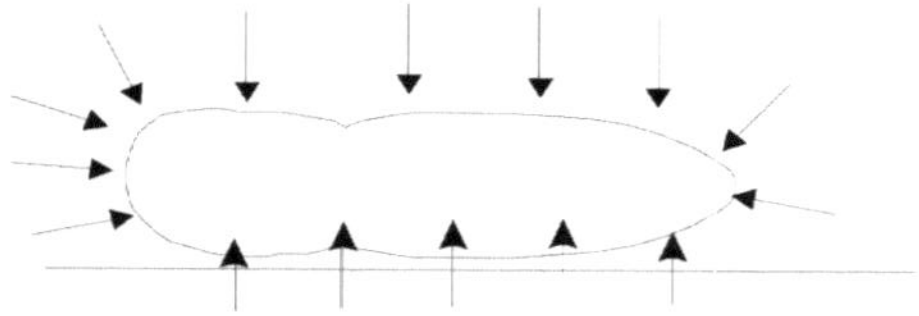

GREATER HEAT TRANSFER RATES REQ. HIGHER TEMP &/OR HIGHER HUMIDITY

SCORCH + HIGH MOISTURE

A CRACKER STORY

GARY ISRAEL
PROCESS SOLUTIONS OF NY LLC
SEPTEMBER 19, 2008

Snack Cracker Production

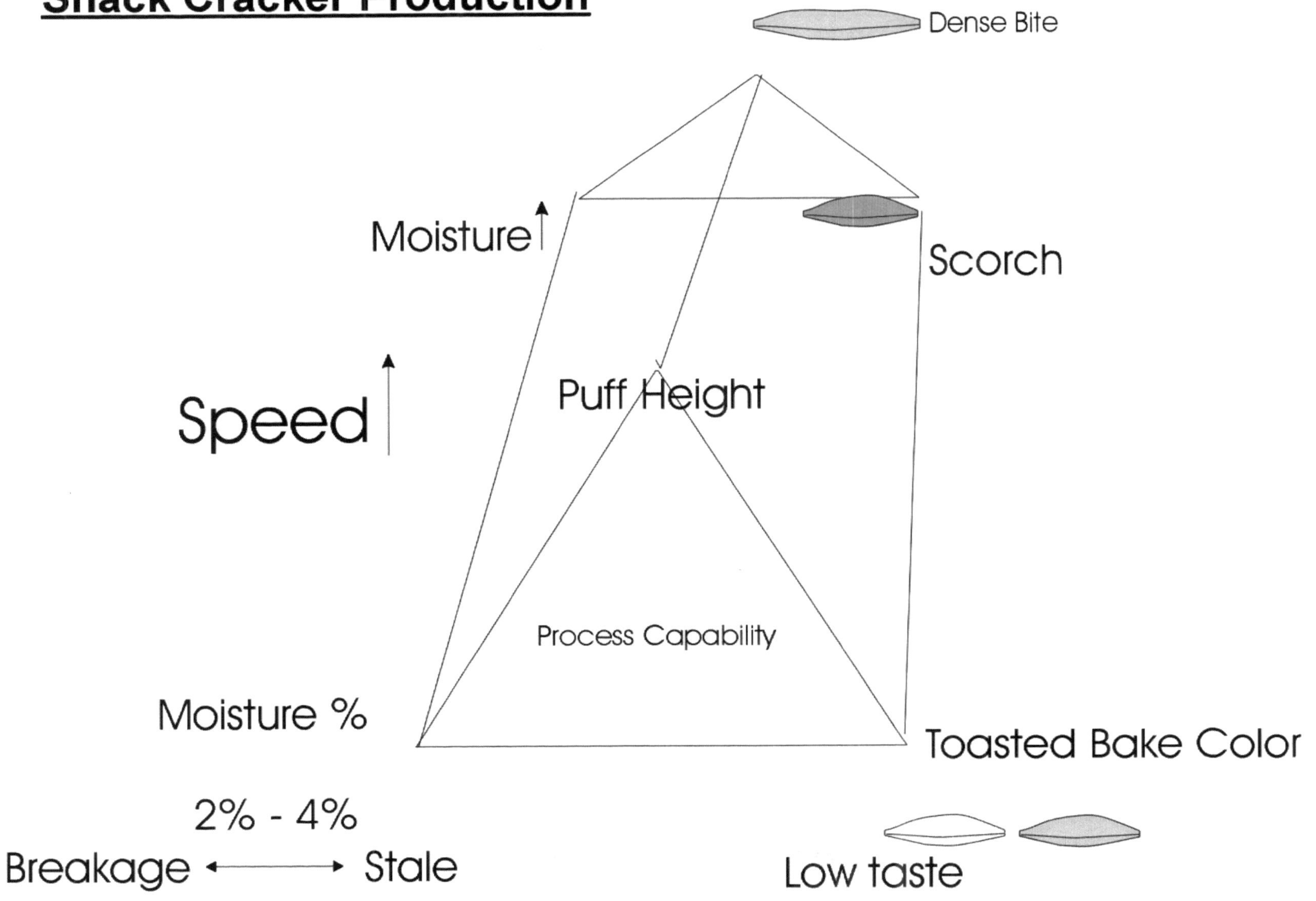

GARY ISRAEL
PROCESS SOLUTIONS OF NY LLC
SEPTEMBER 19, 2008

RADIO FREQUENCY DRYING

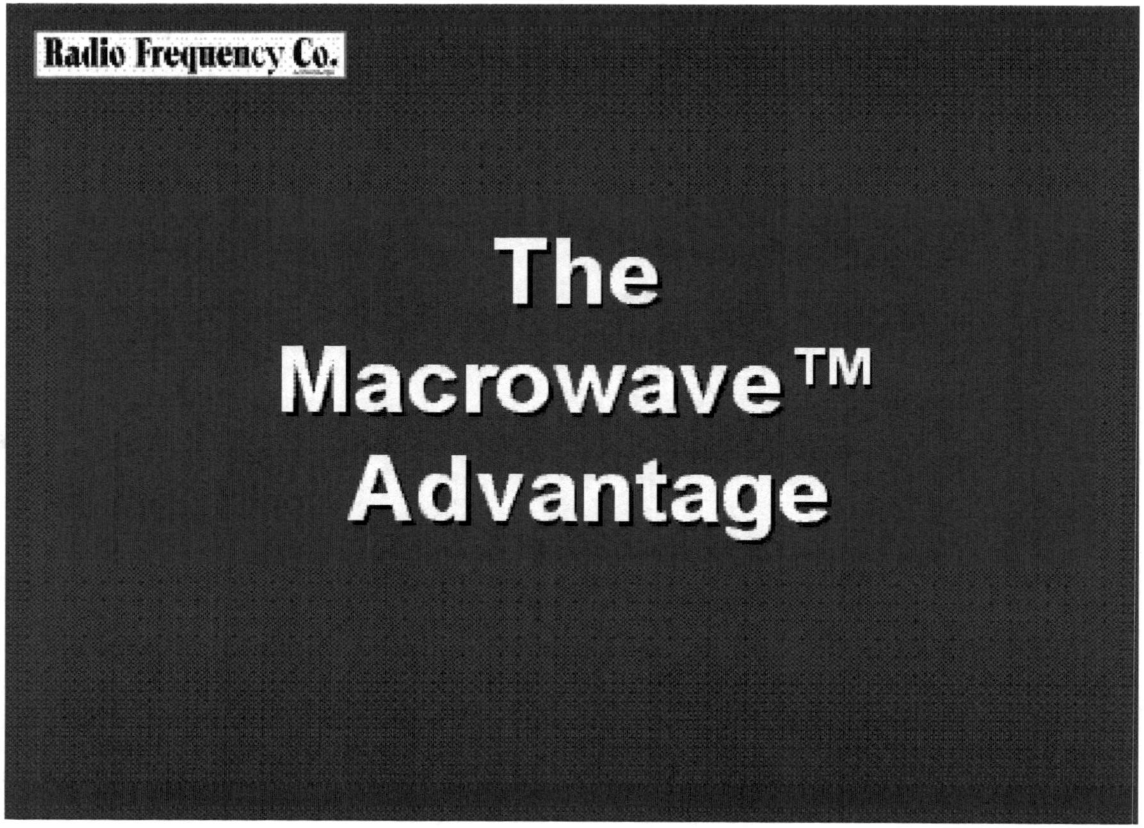

RADIO FREQUENCY DRYING

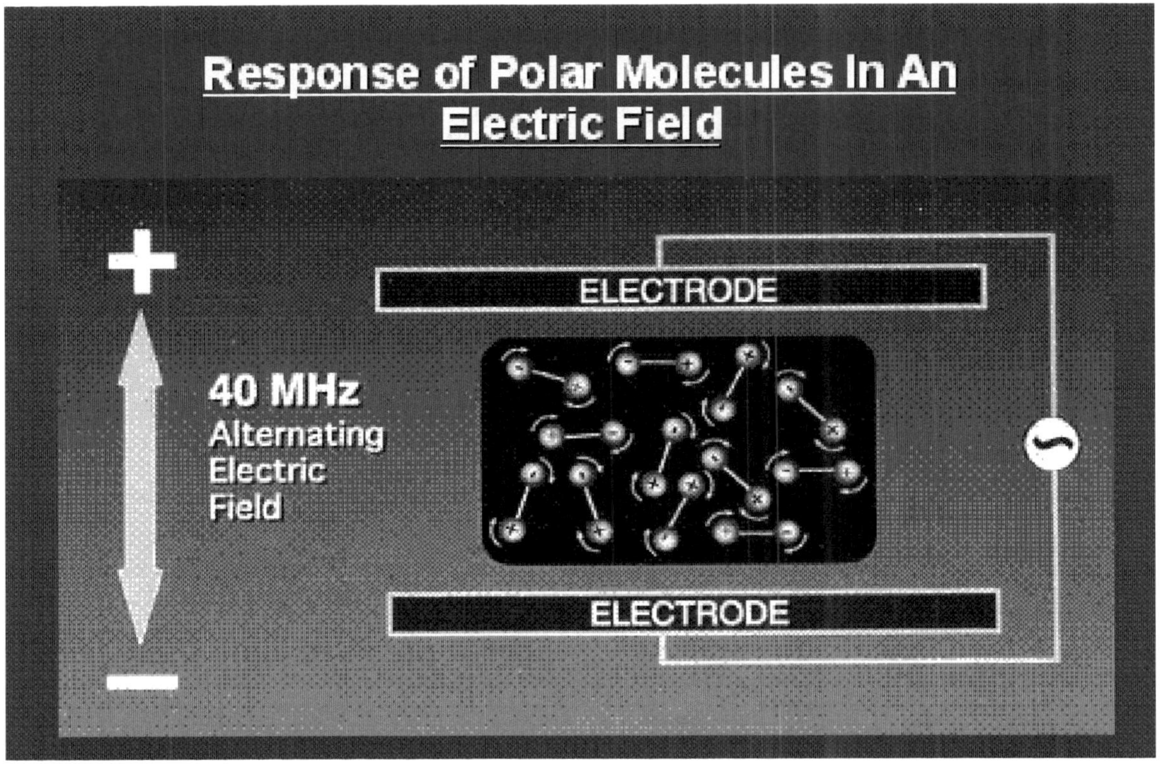

TARGETING POLAR (H_2O) MOLECULES

A CRACKER STORY

GARY ISRAEL
PROCESS SOLUTIONS OF NY LLC
SEPTEMBER 19, 2008

RADIO FREQUENCY DRYING

Dielectric Materials + Electromagnetic Field HEAT

Some portion of the electromagnetic energy will go through a change of state and be dissipated as heat with the dielectric.

Conversion of energy = F (atomic and molecular structure of the material, frequency, field strength)

Alternating electromagnetic field

- *Displacement of the polarized components as re-alignment with the positive and negative oscillations occurs.*
- *Friction on an atomic or molecular level = Heat generated in the dielectric.*

Heat generated in a dielectric

$$P = .555\ f\ E2\ e'\ (\tan \delta \times 10\text{-}6)$$

P = Heat generated in watts/cm3
f = Frequency of the electromagnetic field MHz/sec
E = Field strength in Volts/cm
e' = Dielectric constant of the material
tan δ= Loss tangent, Ie/Ic

A CRACKER STORY

GARY ISRAEL
PROCESS SOLUTIONS OF NY LLC
SEPTEMBER 19, 2008

RADIO FREQUENCY DRYING

Heat generated in a dielectric

$$P = .555\ f\ E2\ e'\ (\tan\delta \times 10\text{-}6)$$

P = Heat generated in watts/cm3

f = Frequency of the electromagnetic field MHz/sec

E = Field strength in Volts/cm

e' = Dielectric constant of the material

tan δ= Loss tangent, Ie/Ic

Materials with higher "loss tangents" and higher "dielectric constants" heat more readily

(Multiply the (loss tangent) by the (dielectric constant) we obtain by definition the "loss factor",

The higher the loss factor of any specific material, the more efficiently it will heat in an alternating RF field.

Materials with loss factors of

- .3 or greater are considered excellent candidates for RF heating,
- between .2 and .3 good candidates,
- <.2 but> .1 fair to poor.

Raw Dough - Bound Water ------ Poor

Dough > 180 F Good !

GARY ISRAEL
PROCESS SOLUTIONS OF NY LLC
SEPTEMBER 19, 2008

RADIO FREQUENCY DRYING

Typical Dielectric Heating Frequencies Reserved For Industrial Use (ISM)

- 13.56MHz + 0.05%
- 27.12MHz + 0.50%
- 40.68MHz + 0.05%

GARY ISRAEL
PROCESS SOLUTIONS OF NY LLC
SEPTEMBER 19, 2008

RADIO FREQUENCY DRYING

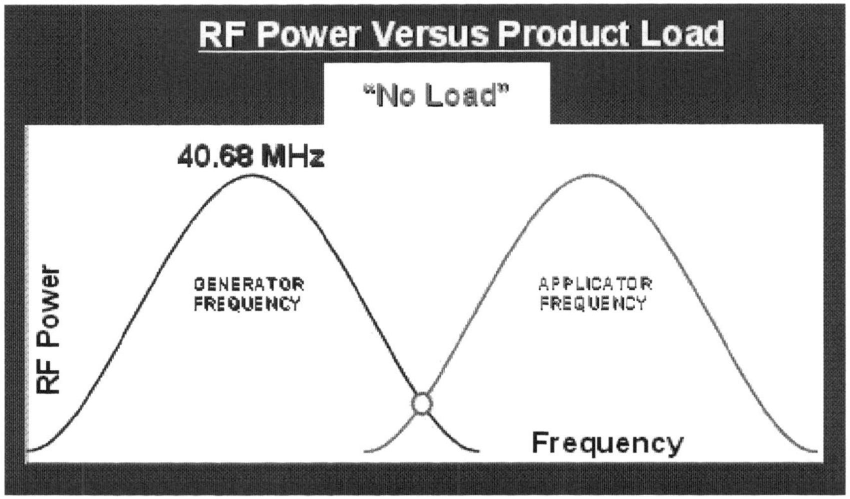

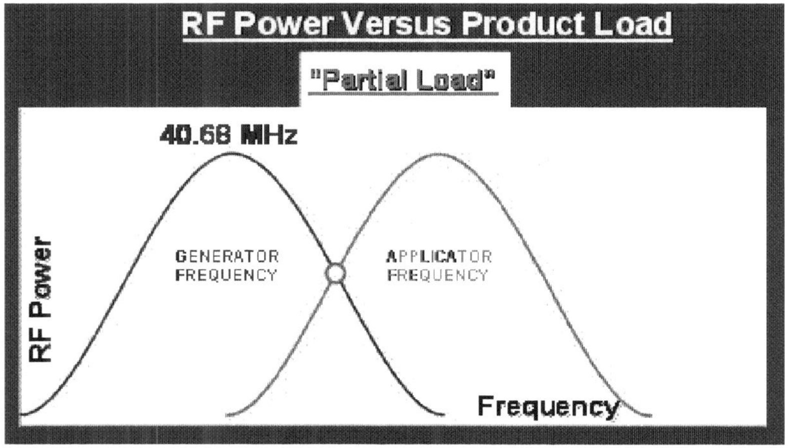

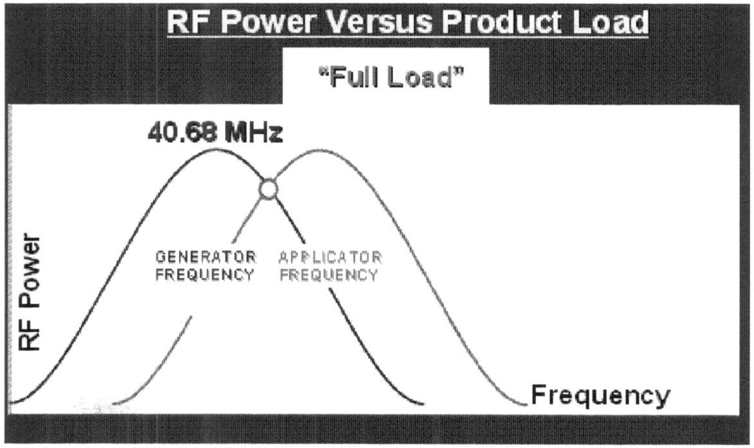

RADIO FREQUENCY DRYING

Radio Frequency Co.

Advantages of Macrowave™ Heating

- Automatic Response to Product Load
  - Power Consumed Proportionate to Product Load
  - Unlike Microwave, there is no need for Energy Wasting Dummy Loads to Accommodate Gaps in Production
  - Inherent Versatility for Multiple Product Lines
- Uniform Application of Heat to the Product Load
  - Longer Wavelength Eliminates Preferential Surface Heating
  - Energy is Applied Uniformly Across Product Width

$$ CAPITOL ADVANTAGES

RADIO FREQUENCY DRYING

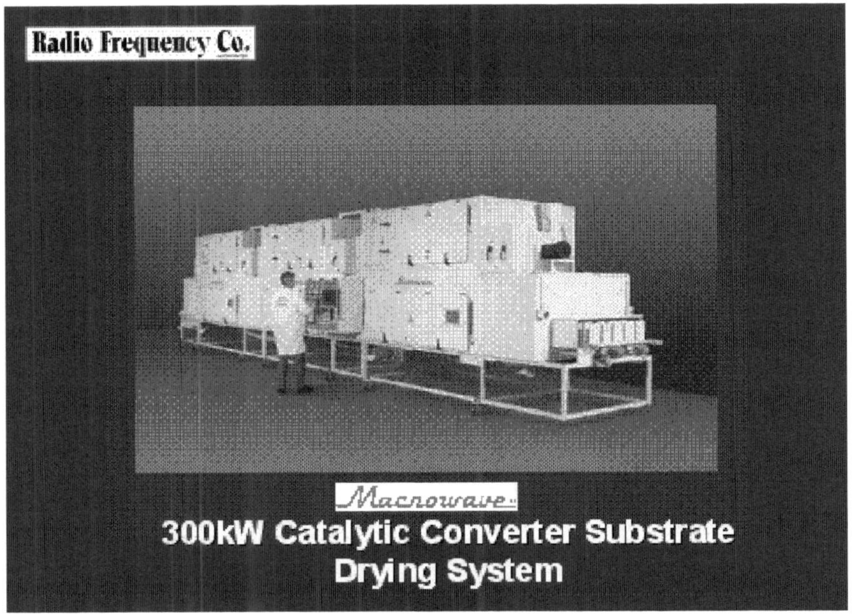

ALTERNATE BAKING OPPORTUNITIES

STANDARD: GDF THERMAL — EXTERNAL HEAT

RADIO FREQUENCY — INTERNAL HEAT

HYBRID OVEN BAKE

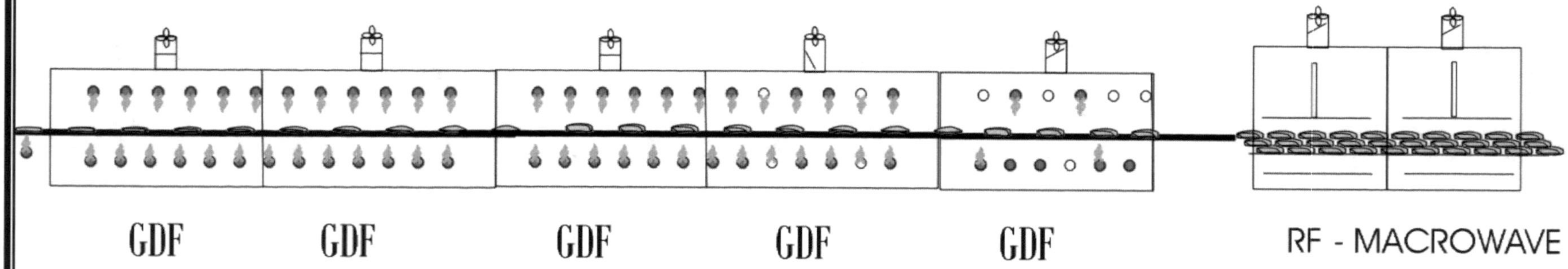

NO SCORCH + MOISTURE CONTROL

GARY ISRAEL
PROCESS SOLUTIONS OF NY LLC
SEPTEMBER 19, 2008

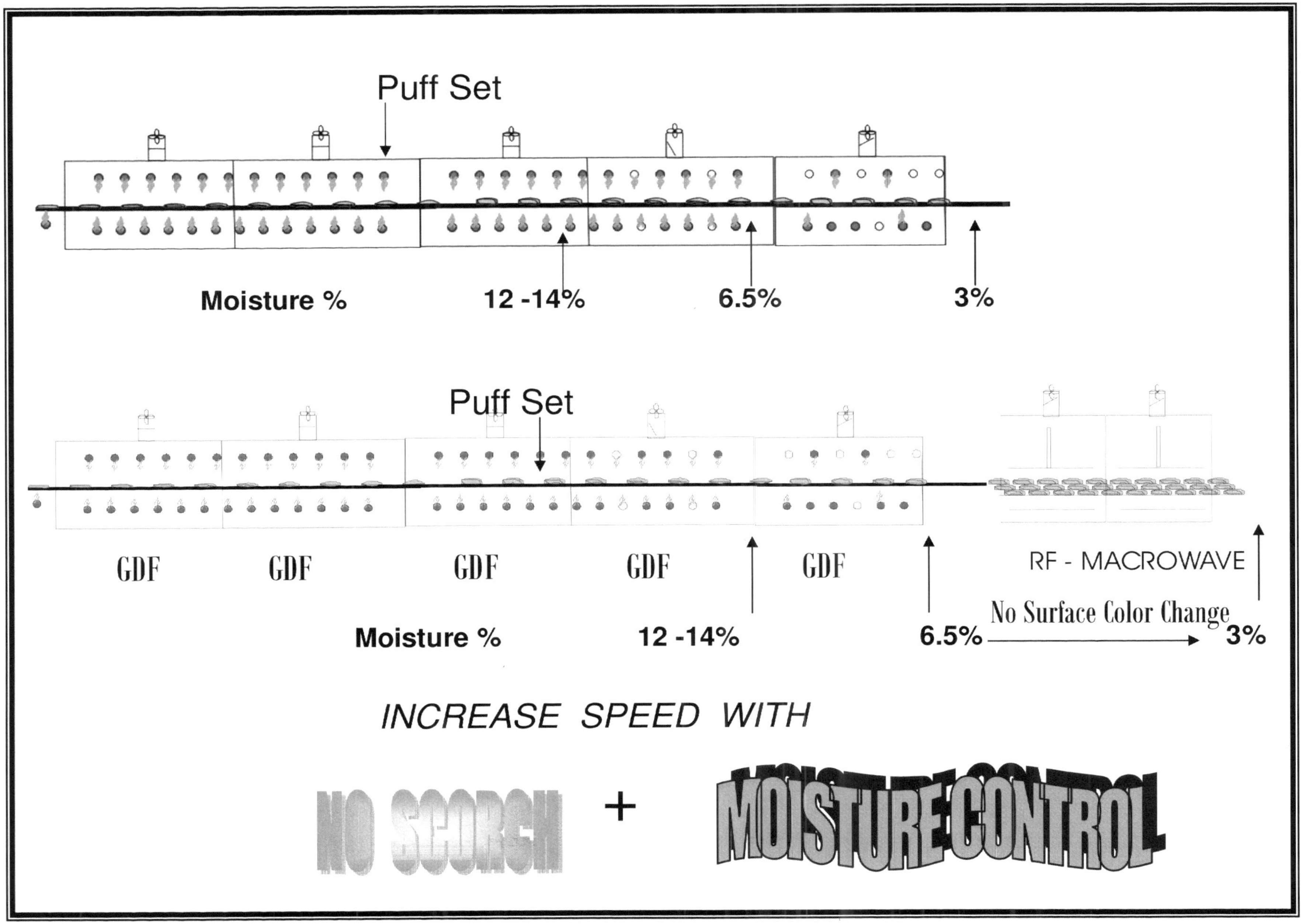

A CRACKER STORY

GARY ISRAEL
PROCESS SOLUTIONS OF NY LLC
SEPTEMBER 19, 2008

DOUBLE THE OUTPUT 2+ YRS

CUTTER OPTIMIZATIONS + HYBRID OVEN BAKE

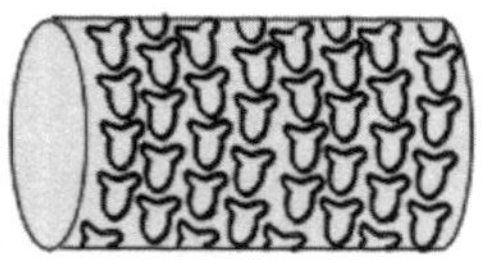

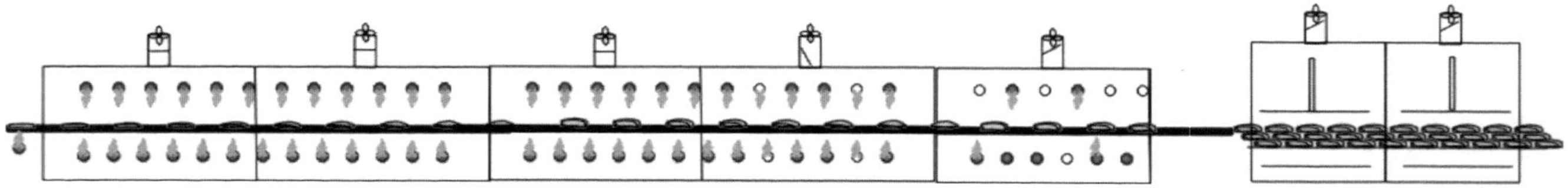

		#/Shaft	RPM	Bake-Min	Crackers / min.	
Project Start	Plastic Cutters	490	30	4.25	14,700	
Stage 1	Brass Cutters	670	30	4.25	20, 100	+37%
Stage 2	Brass Cutters	873	30	4.25	26,190	+78%
Stage 3	Brass Cutters	912	30	4.25	27,360	+86%
Stage 4	Hybrid Bake	912	37	3.44	33,744	+129.%
Stage 5	Hybrid Bake	912	41	3.11	37,392	+154 %
Stage 6	Hybrid Bake	912	49	2.6	44,688	+ 204 %

A CRACKER STORY

GARY ISRAEL
PROCESS SOLUTIONS OF NY LLC
SEPTEMBER 19, 2008

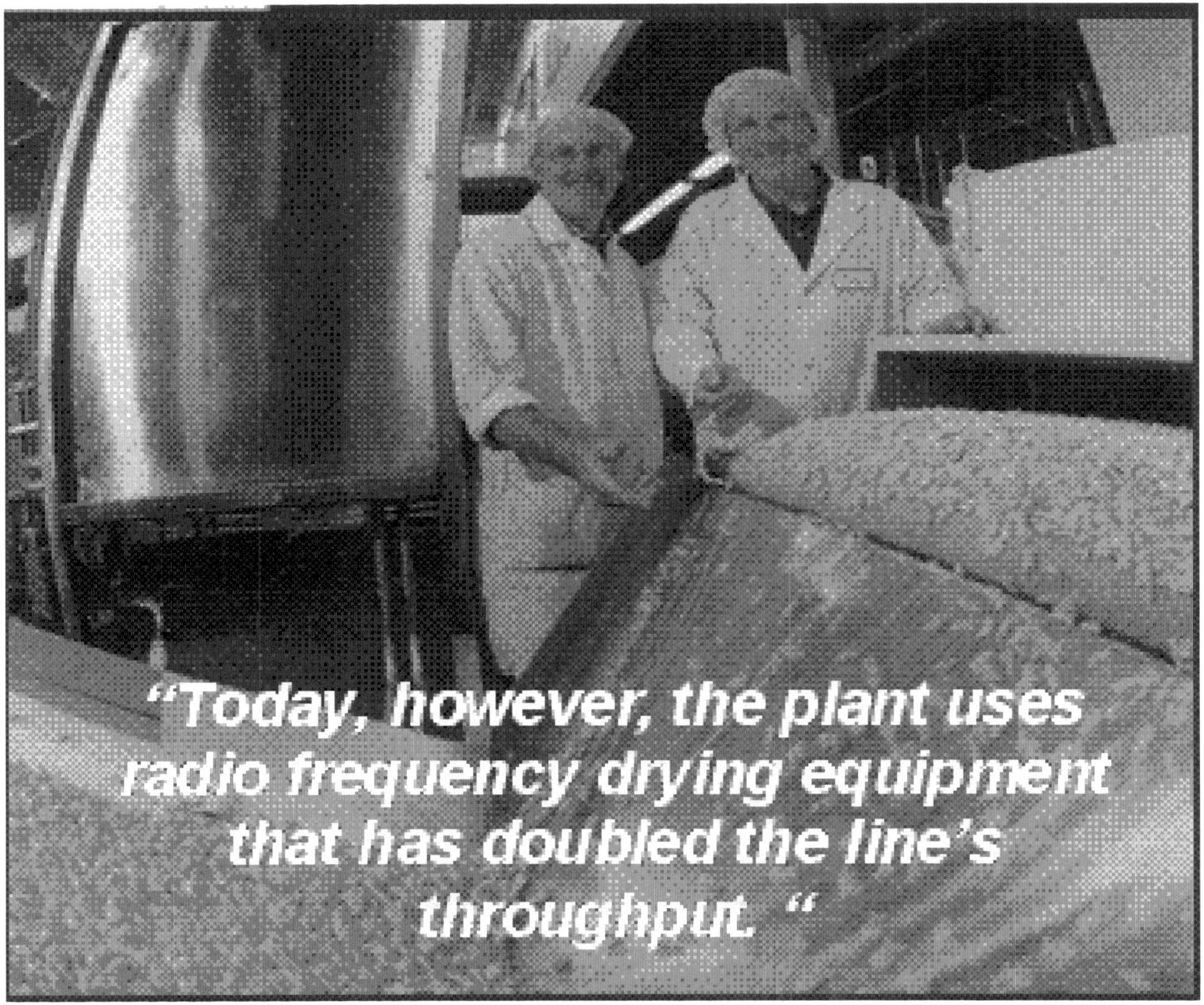
"Today, however, the plant uses
radio frequency drying equipment
that has doubled the line's
throughput. "

632c Product and Process Optimization in Polyester Dyeing

Yue-Hin Leung, Department of Chemical Engineering, The Hong Kong University of Science and Technology, Clear Water Bay, Kowloon, Hong Kong, China and Ka M. Ng, Department of Chemical Engineering, Hong Kong University of Science and Technology, Clear Water Bay, Kowloon, Hong Kong, China.

The dyeing of polyester with disperse dyes has been practiced for the last 60 years. Nevertheless, the dyeing recipe is typically decided by experience based on dye house experts' judgment. This study aims to provide a fundamental understanding of the overall polyester dyeing process, from dye uptake by fibre in a microscopic view, via analysis on the coloration of dyed product, to the operational optimization in the actual process. Through such an understanding, a better process for obtaining a product of the desired quality can be developed. For this, bench scale dyeing experiments were carried out in a special-designed dyeing tank and dye uptake by the fibre were monitored by real time dye solution sampling. Arrhenius and William-Landel-Ferry equations were applied for the changes in dye uptake rate by polyester. Relationship between the three-phase structure for polyester and the dye uptake ability was also studied. Plasticizing effect of one common carrier and solubilizing effect of surfactants were investigated individually. These effects were then related to the color evenness and the luminance of the final product based on the analysis on the final coloration in the CIE L*a*b* Colorspace. Tests on cotton and reactive dyes were also carried out, showing the capability of the experimental setup for various situations. With the understanding on the impact of various effects on the quality of the dyed product, three schemes for process optimization on reducing process time, energy and amount of additives have been proposed and tested.

632d Structured Excipient Design Using Property Clustering and Decomposition Techniques

Charles C. Solvason, Nishanth Chemmangattuvalappil, and Mario R. Eden. Department of Chemical Engineering, Auburn University, 230 Ross Hall, Auburn University, AL 36849-5127

Recent developments in the area of integrated process and product design have shown that products can be designed in terms of their properties with out committing to any specific components a priori. However, the identification of suitable candidate molecules that satisfy a set of property targets is still a challenging issue. While current techniques make use of group contribution methods (GCM) to design molecules, there are many properties which can not be estimated by GCM. Furthermore, not all possible atomic arrangements and structures can be represented in GCM. Hence, there is a need of an efficient methodology for the design of structured molecules. One such approach to structured product design is to combine property clustering with decomposition techniques. This approach first utilizes multivariate characterization techniques to describe a set of representative samples, and then uses decomposition techniques such as principal component analysis (PCA), to find the underlying latent variable models that describe the molecule's properties. The orthogonal nature of these models allows for group-based interpretations and property predictions which can be utilized to design new molecules not found in the original set of molecules. In this work a case study of the design of an excipient for an acetaminophen tablet is presented.

632e Development of a Skin-Care Cream Containing Zinc Oxide Nanoparticles and An Herbal Extract

Yuen S. Cheng[1], Ka W. Lam[1], Ka M. Ng[2], Robert K. M. Ko[3], and Christianto Wibowo[4]. (1) Department of Chemical Engineering, The Hong Kong University of Science and Technology, Clear Water Bay, Hong Kong, Hong Kong, (2) Department of Chemical Engineering, Hong Kong University of Science and Technology, Clear Water Bay, Kowloon, Hong Kong, China, (3) Department of Biochemistry, The Hong Kong University of Science and Technology, Clear Water Bay, Hong Kong, Hong Kong, (4) ClearWaterBay Technology, 4000 W. Valley Blvd. Suite 100, Pomona, CA 91789

The development process of a skin-care cream containing both zinc oxide nanoparticles and an herbal extract, Fructus Schisandrae (FS), is described. The activities regarding product conceptualization and detail design and prototyping are discussed.

In consideration of the market needs, the product is designated a day cream with both sun-screening and anti-oxidant functions. Market survey reveals that customers prefer such a cream product to be non-greasy, soft and with good spreadability. These sensorial and rheological quality factors are translated to corresponding technical specifications, such as emollient content and constituent droplet size. A light cream with low viscosity (6,000-50,000 mPa•s) gives a pleasant spreadability and an oil/water microstructure is preferred due to its non-greasy feel. The cream formula also includes other supporting materials such as emollients, emulsifiers, thickeners, stabilizers, and preservatives, which should be selected to meet the technical specifications and selection criteria. Based on the formula, a prototype cream is fabricated. The prototype is characterized by their pH, droplet size distribution, feel on application, as well as on their stability. If any of the parameters does not meet the specifications, a new prototype is made by adjusting the formula compositions or operating parameters according to a set of heuristics. In addition, a bioassay using rats is established for assessing the biological activity of this anti-oxidant-supplemented sunscreen cream. Significant increases (18-25%) in the amount of natural anti-oxidants, glutathione (GSH) and α-tocopherol (α-TOC), are observed on rat skin treated with FS containing cream by comparing the control cream with no FS. Also, the FS containing cream is found to give protection on the natural anti-oxidant network on the skin against UV irradiation.

632f Stabilizing Copper Hydroxide

Matthew R. Oberholzer, Process Chemistry and Engineering, DuPont Crop Protection, Stine-Haskell Research Center, P.O. Box 30, Newark, DE 19714-0030

Copper hydroxide has been used for many years to combat fungal disease in numerous commercial crops. Copper hydroxide is known to be thermodynamically metastable, and can decompose to copper oxide at elevated temperatures. A simple method was discovered to stabilize copper hydroxide and ensure product stability at elevated temperatures.

717a Computation of Melting Points, Crystalline Free Energies, and Transport Properties of Molten Salts

Saivenkataraman Jayaraman, Chemical and Biomolecular Engineering, University of Notre Dame, 148 Fitzpatrick Hall of Engineering, Notre Dame, IN 46556 and Edward J. Maginn, Department of Chemical and Biomolecular Engineering, University of Notre Dame, 182 Fitzpatrick Hall, Notre Dame, IN 46556.

Ionic liquids, salts that are liquid near ambient temperatures, are finding application in separations, catalysis, and lubrication fields. More traditional salts such as alkali halide eutectic mixtures are being used as industrial heat transfer fluids in solar power generation systems, where it is required to store and utilize heat for producing electrical energy in periods of low or no sun.

Properties such as viscosity, thermal conductivity, density and melting point are all critical for the use of these salts in a given application. Experimental determination of these properties is difficult, particularly for the high temperature systems, and so there is a great desire to develop computational approaches for predicting properties of molten salts.

Melting points have been one of the most difficult properties to estimate from simulations. A new method developed in our group recently [1-2] allows accurate computation of melting points. In our recent study [3], we computed the melting points of the two known crystal polymorphs of the ionic liquid 1-n-butyl-3-methylimidazolium chloride, which are an orthorhombic and a monoclinic structure. We also computed the free energy difference between the two structures, and hence this leads to a new application where we can determine relative stabilities of crystal polymorphs by computing their relative free energy differences.

In the area of heat transfer fluids, we use simulations to predict how properties depend on the ratio of different cations in nitrate-based salts used in solar energy applications.

[1] D. M. Eike, J. F. Brennecke, and E. J. Maginn, J.Chem. Phys. 122, 014115 (2005).

[2] D. M. Eike and E. J. Maginn, J. Chem. Phys. 124,164503 (2006).

[3] S. Jayaraman and E. J. Maginn, J. Chem. Phys. 127, 214504 (2007).

717b Molecular Design of Self-Assembling Copolymer and Surfactant Systems

Shekhar Jain[1], Adam S. Bymaster[1], Chris Emborsky[2], Kenneth R. Cox[2], and **Walter G. Chapman**[2]. (1) Department of Chemical and Biomolecular Engineering, Rice University, 6100 Main St., Houston, TX 77096, (2) Chemical and Biomolecular Engineering Department, MS-362, Rice University, 6100 Main St., Houston, TX 77005

Many novel applications of copolymer and surfactant systems take advantage of their unique microstructure and phase behavior. Example applications include performance polymers, stabilizers for pharmaceutical suspensions, templates for nano-materials, and water-based coatings. The challenge in molecular level design of such systems lies in having predictive models that include the physics that emerges from finite-size effects, varying dimensionality, surface forces and interplay of multiple length and time scales.

In this presentation, we introduce a novel molecular theory for multi-scale modeling of complex fluid assemblies. This computationally efficient classical density functional theory can predict the effects of molecular size, branching, and functional groups on interfacial properties and microstructure. Predictions of the theory are in agreement with molecular simulation results and experiment for phenomena such as polymer depletion, potential of mean force between nano-colloids, and surface-induced segregation – key elements in polymer-colloid systems and in coatings of polymer blends. The presentation will examine how this model combined with other physically intuitive theories can be used as tools of molecular design in copolymer and surfactant systems.

717c A Novel Approach to Process and Product Design Using Molecular Signature

Nishanth Chemmangattuvalappil, Charles C. Solvason, and Mario R. Eden. Department of Chemical Engineering, Auburn University, 230 Ross Hall, Auburn University, AL 36849-5127

Recent developments in the area of integrated process and product design have shown that products can be designed in terms of their properties without committing to any specific components a priori. However, the identification of suitable candidate molecules that satisfy a set of property targets is still a challenging issue. While current techniques make use of group contribution methods (GCM) to design molecules, there are many properties which can not be estimated by GCM. Furthermore, not all possible atomic arrangements are represented in GCM. Hence, there is a need of an efficient methodology for the design of molecules that can handle the multitude of non-GCM described properties. In this work, an algorithm is presented that integrates the concept of molecular signatures into the reverse problem formulation framework. A signature is a systematic coding system of atom types. A molecular signature is a linear combination of its atomic signatures. It has been shown that any of the topological indices (TI) of molecules can be represented in terms of molecular signatures and that it is possible to correlate the TI to the actual properties. Using this information, the new algorithm utilizes the property clustering technique in a reverse problem formulation to obtain the molecular structure based on the property requirements estimated in the process design step. The accuracy of this method depends only on two factors: (1) how well the property-TI relationships are estimated and (2) the level of atomic signatures used to describe the TI. Since, many TI can be used to describe each property, this algorithm generally provide reliable results. This contribution will illustrate the developed methods and demonstrate their use in a case study.

Computer Aided Polymer Design Using Multiscale Modelling

Kavitha C. Satyanarayana[1], Jens Abildskov[1], Rafiqul Gani[1], Georgia Tsolou[2], Vlasis G. Mavrantzas[2]*

[1]CAPEC, Department of Chemical Engineering, Technical University of Denmark, DK-2800 Lyngby, Denmark.

[2]Department of Chemical Engineering, University of Patras & Institute of Chemical Engineering and High-Temperature Chemical Processes (ICE-HT/FORTH), Patras GR 26504 Greece

kac@kt.dtu.dk, ja@kt.dtu.dk, rag@kt.dtu.dk, georgia@iceht.forth.gr, vlasis@iceht.forth.gr

Abstract: Different approaches are being investigated to design a new product with the required properties. These can be broadly classified under different scale depending upon the length and time of the approach investigated. An integration of macro to micro scale for designing a polymer, with the required property, is made in this work. Computer Aided Molecular design (CAMD) algorithm, which uses group contribution plus (GC^+) models for predicting polymer repeat unit properties is classified under macro and meso scale approaches. It is employed in this work to identify polymer repeat units matching a set of desired properties. The identified repeat unit structures from the CAMD algorithm is arranged to form a polymer chain and its structural and conformational properties are studied using an atomistic simulation that is classified as a micro-scale approach. This integration of multiscale approaches can decrease the time of developing new polymers with the desired property as it can readily

[*] Corresponding author Tel. +45 4525 2882, Fax +45 45932906, email rag@kt.dtu.dk

forecast the polymer structures that has the required set of properties. A detailed case study using this multiscale methodology is presented in this paper.

KEYWORDS – CAMD, polymer design, molecular dynamics, group contribution plus methods, polyisobutylene, diffusivity.

INTRODUCTION

Polymers are mostly used in chemicals based consumer products in some form. The demand for new polymer based products is increasing day by day. The search for new polymer products with desired end-user properties has become a primary objective in chemical, pharmaceutical and food industries. Traditional approach of preparing and testing many candidate polymers at the laboratory scale, to verify if they have the required properties is well known. Though this approach can be accurate enough to discover a new polymer, yet it is highly time consuming, tedious and capital intensive.

There are several other approaches, which can be used to forecast the polymer repeat unit structures that posses the required properties, prior to the traditional synthesis-and-test approach. Some of these approaches are shown in the figure I. These approaches are unique in their own way but do have limitations on their applications. At times, these approaches can also be complimentary to each other. For example, in Computer Aided Molecular Design (CAMD) solution steps, the property constraints are evaluated using a set of pre-selected predictive property models based on groups, or atom-connectivity; and molecular descriptors to represent the generated molecular structures at different size scales and structural details. This requires a multiscale modeling approach involving the selection of repeat unit (at macroscopic scale) and its configurational arrangement (at microscopic scale).

Identification of a basic polymer repeat unit at macro scale level requires an appropriate CAMD algorithm and a set of property prediction models. In the CAMD algorithm, fragments are used to design the homo polymer repeat unit structures. These can either be groups at macro-scale or atoms at meso-scale or both; and property prediction models are based on these types fragments *(1-3)*. Therefore, the CAMD algorithm completely depends on the reliability of the property prediction model used.

Group contribution based property prediction models *(1,3)* are sufficiently simple and reliable for use in CAMD, but due to the limited availability of experimental data for polymer properties, one cannot expect to fill out the entire group parameter table. At times there can also be a polymer repeat unit structure, which cannot be totally represented by groups for a specific group contribution method. This is overcome by a meso-scale approach, where an atom-connectivity index method is applied to determine the missing groups and/or the missing group contributions automatically without the need for additional experimental data. This integration of the group and atom-connectivity index based models is called as group contribution plus (GC^{+}) model and it has been developed for prediction of properties of polymer repeat units by Satyanarayana et al. *(3)*. Since, the atom-connectivity index based property prediction models have comparatively lower accuracy than the group contribution models *(4)*, they are used to fill out the gaps in the group contribution models instead of relying totally on them. After the integration of macro and meso-scales for the fragments to be used in the CAMD algorithm, the next step is to decide which type of CAMD algorithm can be used.

Different types of CAMD algorithms have been proposed for homo polymer repeat unit structure identification for a specified (desired) set of property criteria. Some of them being a mixed integer linear programming (MILP) based algorithm using group contribution property models by Maranas *(5)*; a genetic algorithm based CAMD-algorithm using group contribution property models by Venkatasubramanian et al. *(6)* and a multi-step, multi-level hybrid CAMD algorithm by Harper and Gani *(8)*. A multi-step, multi-level hybrid CAMD algorithm developed by Harper and Gani *(8)* suppresses the combinatorial explosion problem and also generates a large number of structures. This method has been modified for using GC^{+} models by Satyanarayana et al *(7)* and it is used in this work.

Configurational arrangement of the identified basic polymer repeat unit can be done by using molecular mechanics at micro scale level and these arranged structures can later be used for either prediction or validation of the required properties using atomistic simulations (micro scale). This approach depends on reliability of the model used for force fields and also on the accuracy of the conformational and structural properties of polymer chain. These simulations are computationally intensive and time consuming, as such, this can be used immediately after the basic polymer repeat unit structure has been identified using the CAMD algorithm. Some atomistic simulation based studies to predict the property of polymers have been done by Makrodimitri et al. *(9)*, Tsolou et al. *(10)*, Charati and Stern *(11)*, Boyd and Pant *(12)*, Müller-Plathe et al *(13),* etc.

By integrating the different scales, from macro- to micro-scale, the uniqueness of each method is retained and their limitations are avoided to some extent. Integration of multiscale approaches and using this to design a polymer repeat unit structure (as well as its chain length distribution), meeting a set of target properties is the main goal of this paper. The method and its application is highlighted through a case study.

BACKGROUND

Group Contribution plus (GC^+) model – This property prediction model is the integration of group contribution method with atom-connectivity index method developed by Gani et al. *(4),* with emphasis on properties of organic chemicals. It has been further extended for predicting homo-polymer repeat unit properties in this work using Marrero/Gani group contribution method and atom-connectivity index method *(3)*. Marrero/Gani group contribution method *(14)* has advantages over other group contribution methods due to the large range of groups, classified as first-, second- and third order, which may be used in models for polymer property prediction as well as for generation of polymer-repeat unit structures.

The two main requirements of any group contribution method are:

(1) To represent the polymer repeat unit structure completely with the groups available in the selected property prediction method.

(2) To estimate the necessary property by using the property contributions of the groups used to represent the polymer structures.

If the polymer repeat unit structure of *poly(bismethoxyphosphazene) (15)* is considered (see figure II), Marrero/Gani group contribution model cannot fully represent this structure; and hence its properties cannot be estimated by this method. Moreover, as the number of reliable data on polymer systems is limited, one cannot expect group contribution tables covering large ranges of groups without missing entries. Thus, such a method alone may not be very widely applicable because the generation of missing group contributions cannot be totally ruled out.

To retain the advantages of Marrero/Gani group contribution methods and to overcome the limitation of their missing contributions, this group contribution method is integrated with the atom-connectivity index method, which is called GC^+ model.

Marrero/Gani group Contribution method

The Marrero/Gani group contribution model is usually written as follows:

$$f(X) = \sum_i N_i C_i + w \sum_j M_j D_j + z \sum_k O_k E_k \qquad \text{...(Eq. 1)}$$

Where, C_i is the contribution of the first-order group of type-i that occurs N_i times, D_j and E_k are the contributions of the second-order group of type-j and the third-order group of type-k, that occurs M_j and O_k times, respectively. In the first level of estimation, where $w = z = 0$, only first-order groups are employed. In the second level, where $w = 1$ and $z = 0$, only first- and second-order groups are involved. In the third level, both $w = 1$ and $z = 1$, and contributions of groups of all levels are included in the calculation. The left-hand side of Eq. (1) is a simple function, $f(X)$, of the property, X. When selecting this function one tries – as closely as possible - to satisfy the following criteria:

(1) The function has to achieve additivity in the contributions C_i, D_j and E_k (thereby providing linear programming problem formulation for CAMD)

(2) It has to enable the best possible fit of the experimental data (thereby providing the necessary accuracy)

(3) It should provide good extrapolating capability and therefore, a wide applicability (thereby providing reliability of innovative solutions)

Thus the procedure for property prediction is the same as in the case of organic chemicals, except that here the same set of groups describes the polymer repeat unit structures, that is, structures having two free attachments.

Atom-Connectivity index method

In this case, the polymer repeat unit structures are represented by atoms and atom-connectivity indices, as proposed by Gani et al. *(4)*:

$$f(Y) = \sum_i (a_i A_i) + b({}^v\chi^0) + 2c({}^v\chi^1) + d \qquad \text{...(Eq. 2)}$$

Where, Y is the sought polymer property, A_i is the contribution of atom i occurring in the molecular structure a_i times. ${}^v\chi^0$ and ${}^v\chi^1$ are the zeroth- (atom) and first-order (bond) connectivity indices, respectively, as described by Kier and Hall *(23)*. b, c and d are adjustable parameters. These models have very few adjustable parameters: A_i (one per atom i), b, c and d.

Development of GC^+ models for predicting polymer properties are explained in detail in Satyanarayana et al *(3)*.

CAMD algorithm for Polymer repeat unit design

CAMD algorithm for polymer product design has three main steps: *problem formulation, repeat unit identification* and *result analysis*, similar to the work by Harper and Gani *(16)*.

Problem formulation

The main aim of using a CAMD algorithm is to identify a set of candidates or compounds possessing a specified set of properties. This can be achieved by generating molecular structures that match the required specifications with respect to compound type, and physical and chemical properties. In order to find a set of polymer repeat unit structures having desired properties, it is necessary to provide a set of specifications for these (target) properties. The molecular design problem may be formulated to have only structural constraints or only property constraints, or both, depending on the specific needs for the desired product.

Structural constraints – these are related to the size and chemical feasibility of the generated structures. For example, for a representation by groups, how many groups (maximum and minimum) can be present, how many times a particular group can occur in the repeat unit, and the rules for combining them? Moreover any chemical structure that is generated should also satisfy certain structural conditions, such as; there must be two free attachments in each polymer repeat unit structure. The combination rules have been adapted from those of Harper and Gani *(16)*.

Property constraints – these are related to the needs for the desired product, expressed in terms of target property values. Mathematically, these constraints are written as $P_l \leq P(n_i) \leq P_u$, where n_i is the number of times the *i*th group appear in the polymer, $P(n_i)$ is the vector of predicted properties for the polymer, P_l is the vector of lower bounds specified on the various polymer properties and P_u is the vector of upper bounds specified on the various polymer properties.

Repeat unit identification

The employed CAMD technique is of the "generate-and-test" type, where all feasible repeat unit structures are generated from a set of building blocks and subsequently tested against the property constraints. The multi-level approach of Harper et al. *(8)* is employed, basically to avoid a combinatorial explosion. This multi-step, multi-level hybrid CAMD algorithm is modified for polymer repeat unit design, which uses GC^+ model for predicting property of the generated polymer repeat unit structures *(19)*. Hence, polymer repeat units are generated in successive steps, where they are screened against a corresponding set of property constraints. The level of molecular details increases as the number of promising candidates decreases, in each subsequent step.

Result analysis

The screened repeat units satisfying the property constraints can be further analyzed; if necessary, using literature sources or other prediction methods, as additional design steps.

Molecular Dynamics

The identified basic polymer repeat unit from the CAMD algorithm does not give the complete structural or conformational information of the polymer. More over it is not one basic repeat unit that constitutes the whole polymer, but number of basic repeat units that together form a characteristic polymer. It is therefore required to arrange the identified basic repeat unit structures in such a way that it has full conformational information of how a polymer repeat unit could exist in reality. This is a missing part of the CAMD algorithm but it can be complimented by using molecular dynamics to arrange the polymer repeat unit structures and obtain the conformational information and verify the chain-length for the polymer. Having the available structural information, the polymer repeat units can be set for atomistic simulations to predict and/or validate how the property corresponding to these structures change as a function of size (number of repeat units) or operational variables such as, temperature. These atomistic simulations are done in this work using Material Studio, software package of Accelrys Software Inc. *(17)*.

METHODOLOGY

The schematic work flow of multilevel modeling for predicting polymer repeat unit structure with the required target properties is shown in the figure III. The required target properties are set as the input to this multilevel modeling methodology. The modified CAMD algorithm *(7)* using either macro or meso scale approach or both, is used generate the basic polymer repeat unit structures and check if they posses the required property or not (using GC^+ model). A set of basic polymer repeat unit structures which promises to have the required properties are obtained as output from this step.

A number of basic polymer repeat units, which are identified from macro and meso scale approach, are arranged to form a homo-polymer repeat unit structures. The united-atom force fields model, which incorporates a group of atoms that can approximately represent the molecular mechanical properties of the group on a scale of size that is larger than atomic scale, is used to describe these structures. The united-atom model is a good approximation to simulate molecular systems in which the intermolecular motion is much more important than the intramolecular motion. These polymer repeat unit structures are simulated extensively to determine or validate conformational characteristics and volumetric properties over a wide range of temperatures. This extensive simulation is done in stage A as described in the *micro scale level* part of the figure III. The independent configurations of the polymer obtained from stage A is used in stage B, as shown in figure III. These pre-equilibrated configurations with united-atom model are converted to all-atom (explicit-atom) structures, re-equilibrated using the detailed COMPASS (Condensed-phase Optimized Molecular Potentials for Atomistic Simulation Studies) force field and then simulated shortly to estimate or validate the required property. The conversion to all atom model is an essential step for the prediction of properties (like barrier properties) because hydrogen atoms in the polymer matrix exhibit enhanced mobility there by causing an increase in the dynamic flexibility of the system through the fast local re-arrangements of the accessible accessible volume. The conversions from a united-atom to all-atom representation do give rise to steric overlaps, but these can be easily eliminated by subjecting the system to an energy minimization procedure.

The obtained polymer configurations that have the required set of properties can then be tested in laboratory to verify if they can be synthesized and also to validate if they have the required property. This methodology can reduce the need for laboratory scale synthesize-and-test of thousands of polymers to verify if they have the required properties, thereby, reducing the time and cost for developing a new polymer based product.

CASE STUDY

A hermetic seal based stopper has wide applications in food, chemical and pharmaceutical industries, as these seals intend to secure against the entry of air or microorganisms and to maintain the safety and quality of the contents stored in the containers with this seal. There are a number of materials that are used as sealants such as natural rubber, cork, etc. Natural rubber has a disadvantage to be used as a stopper for medical (or drug) bottles as they may cause latex allergies on patients *(24)*. Cork stoppers also have disadvantages as they may at times spoil the contents stored with this stopper due to their varying pores size *(25)*. It would be interesting to design a polymeric material which can be used as a hermetic stopper (sealant) by retaining the advantages of the existing stopper materials and overcoming their disadvantages. The multilevel modeling methodology can be used for designing this polymeric material that can be used as a hermetic stopper. But designing a hermetic seal in common to be used in the food, chemical and pharmaceutical industries could be too ambitious, as each of these areas may have different types materials needed to be stored. Therefore, narrowing the problem definition, this case study is confined to the design of a polymeric material that is suitable for use as a hermetic seal in pharmaceutical industries. In other words, it is required to design a polymeric material that can be used as a medical container stopper, which is a hermetic sealant.

Application of the Macro-Meso-scale approach

The design problem to be solved through this methodology is to identify the basic polymer repeat unit structures that can be used as a hermetic stopper, using the macro and meso scale approach. These require a set of structural and property constraints, which are explained in detail below, need to be specified first.

Property Constraints

Glass transition temperature: The polymeric stopper is required to be elastic over a wide range of temperature, and it is preferred to have a rubbery amorphous polymer; which implies that the glass transition temperature of the polymer should be less than or equal to the room temperature,

300 K *(1)*. The upper limit for glass transition temperature is relaxed to 350 K. By doing this, the scope of missing some candidates that have the required property (but are not identified due to inconsistency of the experimental data reported and also considered while developing the property prediction models) is minimized to some extent.

Dielectric constant: Some contents stored in the medical containers needs to be constantly agitated with electrical agitators before use. It is therefore required that the polymeric stopper that is required to be designed should not conduct electricity and affect the stored contents in the medical containers. This implies that the polymer designed should have low dielectric constant. Therefore, maximum constraint can be imposed as 2.5 for dielectric constant.

Low permeation to moisture and air: The focus of this case study is to design a stopper for medical containers that is a hermetic sealant. This implies that the stopper should protect the contents stored from moisture and air; that means that the designed material should have very low water (or moisture) absorptivity *(1)*.

Low permeation to air is also an important aspect to be considered. Air is a mixture of nitrogen and oxygen in larger proportions when compared to other gases comprising it. Low permeation of air through the polymeric stopper that needs to be designed implies low permeation to nitrogen (N_2) and oxygen (O_2). Since there is no direct group contribution method to study the solubility and diffusivity (the key factors to determine permeation) in the polymer, this can be studied at micro scale after identifying the basic polymer repeat units that can be used as a hermetic stopper.

Structural Constraints

Based on the property constraints, it is more efficient to pre-select a basis set of groups (or building blocks) that are needed to build the basic polymer repeat unit structures. In this way, the combinatorial size of the problem is reduced and redundant structures are eliminated easily. Since it is required to design rubbery amorphous polymer (as stated above) with low permeability to moisture, the polar and hydrogen bond forming functional groups are not included in the basis set. The reason for this is, the presence of polar and hydrogen bonding functional groups on polymeric chain often enhances chain-chain attractions (as they make possible dipole-dipole hydrogen bonding intermolecular forces), particularly if these involve hydrogen bonding and there by enhancing crystallinity and tensile strength.

Considering that the polymer repeat unit does not contain any polar or hydrogen bonding forming groups (also aromatic groups), only the groups forming olefinic polymers can be allowed. It is also known that the double bonds in polymers make them poor for weather resistance, thermal resistance, oxygen resistance and ozone resistance, thereby limiting their influence *(18)*. This applies to olefinic groups with triple bonds. Taking this into account, olefinic groups with double and triple bonds are not included in the basis set for designing a polymer repeat unit structure that can be used as a stopper for medical containers. Therefore the basis set of groups included in this case study are CH_3, CH_2, CH and C.

Upper and lower bounds on occurrences of groups: Apart from deciding the basis set, it is also required to decide the minimum and maximum number of groups that can occur in a basic polymer repeat unit structure and also the number of time each group can be repeated. As the side chain forming groups like CH and C are included in the basis set, it is safer to avoid too many side chains in a basic polymer repeat unit and also the longer side chains (in the combination with CH_2 and CH_3). Therefore, the number of times each group can occur in a repeat unit is set to 2. The minimum number of groups' occurring in the basic repeat unit is set

to 2. Since the maximum number of times each group can occur is set to 2, the maximum number of groups that can occur with the combination of all the groups that are included in the basis set comes to 6. Thus;

$$2 \leq \sum_j n_j \leq 6 \qquad \text{...(Eq. 3)}$$

where, n_j is the number of groups of type j.

The structural groups considered in the basis set were classified on the basis of their valency (number of free attachments) and general octet rule is followed while generating structures using these building blocks. As this case study aims at generating a basic polymer (homo) repeat unit that has two free ends, this criterion of having two free ends in each generated structure is also specified. Thus a simple relation for structural feasibility of a collection of groups with this constraint is;

$$\sum_j (2-v_j)n_j = 2(m-1) \qquad \text{...(Eq. 4)}$$

where n_j, v_j are the number and valency, respectively, of groups of type j and m =1, 0 or -1 for acyclic, monocyclic and bicyclic groups, respectively. Here in thos case, as only olifinic groups are considered only acyclic polymer repeat units are generated. Therefore m takes the value of 1.

Structural and property constraints are given as the input to the CAMD algorithm. 112 candidates of basic repeat unit structures are designed and 106 candidates among them are selected based on the given constraints. These 106 candidates include also some isomers of the generated repeat units. Removing the isomers, 12 basic polymer repeat units that satisfy all constraints are listed in table 1.

Repeat unit structure of polyisobutylene is one of the identified repeat units from the CAMD algorithm. Polyisobutylene is an important (known) elastomer with a low glass transition temperature. From different literature sources *(12, 13, 19)* it is reported that this polymer is well known for its markedly low permeability properties to small gas molecules, when compared to other elastomers, due to the presence of two bulky pendant methyl groups on each basic repeat unit structure. Permeability is described as the product of solubility and diffusivity. Lower solubility and diffusivity values always correspond to low permeability. Therefore, solubility and diffusivity of the gas molecules (O_2 and N_2) can be verified or studied under the micro-scale approach. As it is highly time consuming to study the entire identified basic polymer repeat unit structures at micro-scale level (at this point of time), only polyisobutylene is studied at micro-scale level in this paper.

Application of the Micro-scale approach

Stage I

As shown in figure III, this stage deals with the structural representation of polymeric systems. Different polyisobutylene (PIB) systems at various temperatures starting from 8 basic repeat units to 80 basic repeat units to represent a reasonable distribution of chain lengths were studied by Tsolou et al *(10)*. In this study included a brute-force application of a parallel molecular dynamics simulations based on a new united-atom model. Resuts from this work by Tsolou et al *(10)* correctly reproduces the volumetric, structural and conformational properties of PIB. The correct reproduction of the polymeric system is very important to make further investigations on properties. Therefore, these structures have been taken in stage 2 to study the solubility and diffusivity of O_2 in PIB systems.

Stage II

This stage deals with the further investigation of properties from the well represented structures from stage I (see figure III). Fully equilibrated configurations from the simulations performed by Tsolou et al. *(10)* were used in the prediction of solubility of O_2 and N_2. These two molecules have lower solubility coefficient in PIB system as reported by Tsolou et al. *(10)*. As solubility alone is not sufficient to describe permeability, diffusivity of these molecules in the same PIB system needs to be studied. Since the molecular diameter of N_2 is larger than O_2, it is reasonable to just study the diffusivity of O_2 alone as lower diffusivity of O_2 implies lower diffusivity of N_2 *(22)*.

Description of PIB system used in this work: Independent configurations of PIB system of 80 basic repeat units (a total of 320 Carbon atoms in each chain) to represent each chain is considered for studying the diffusivity of O_2. If a number of shorter chain lengths are considered in the simulation box, it can result in the increase in the free volume there by lowering the density of the system (see figure IV). Inorder to have the density of the system closer to the e experimental density, longer chains are taken in the simulation box.

Step wise approach of stage 2: A small simulation box with three chains of PIB-C320 which is thoroughly pre-equilibrated is taken at 350, 450, 500 and 550 K. This system is then converted from united- to all-atom representation using Material Studio *(17)*. The energy of the system is minimized. 5 O_2 molecules are inserted randomly at different positions to each PIB-C320 matrix at different temperatures and the energy of these matrixes are re-minimized. These systems are further thermally re-equilibrated through a short NVT molecular dynamics run for a short time. Later they are simulated through NPT runs (up 2 ns depending on the lower temperature considered).

Trajectory Analysis to calculate diffusivity of O_2: The mean square displacement (MSD) of the gas molecules is calculated and averaged over number of molecules from the trajectory saved during the molecular dynamics simulations. MSD versus time, which is a straight line, is plotted and the slope gives the diffusion coefficient. Diffusion of small molecules in polymer using molecular dynamics simulation is most frequently calculated by means of the Einstein relationship which is given as *(20)*:

$$D_\alpha = \frac{1}{6N_\alpha} \lim_{t\to\infty} \frac{d}{dt} \sum_{i=1}^{N_\alpha} \left\langle \left[r_i(t) - r_i(0) \right]^2 \right\rangle \qquad \text{...(eq. 3)}$$

where, r_i denotes the position vector of center of mass of species α, and angular brackets denote averaging over all choices of time origin within a dynamic trajectory. From an molecular dynamics run, long enough to ensure that the system has reached the fickian diffusive regime (when slope of the log of MSD averaged over number of atoms/molecules *(N_α)* is plotted against time. A plot of MSD as a function of time should give a straight line, if the molecular motion is described by the diffusion equations, and the diffusion coefficient may be determined from the slope of the line *(21)*. MSD versus time for O_2 in PIB-C320 at 350, 450, 500 and 550 K are shown in chart I.

Diffusion Coefficient of O_2 in PIB C-320: The diffusion coefficient of O_2 molecules calculated at different temperatures is shown in chart II (log (D_0) vs 1/T). From this chart the diffusion coefficient of O_2 at 300 K in PIB-C320 can be extrapolated. This value comes to 0.021 Å^2/ps which is two orders higher in dimension to that reported in Müller-Plathe et al. *(13)*. Activation energy for the diffusion can also be calculated using the slope from the chart II in the equation,

$$E_a^{app} = -\frac{R}{0.434}\frac{\partial(\log D)}{\partial\left(\frac{1}{T}\right)} \qquad \ldots\text{(eq. 4)}$$

The calculated activation energy using equation 4 is 5.7kcal/mol and that reported by Amerongen *(22)* is 11.9kcal/mol. Moreover, from the same literature it has been found that activation energy increases as temperature decreases. Consequently the value of 5.7kcal/mol for activation energy is extracted from this simulation, is quite reasonable.

The reason for why the obtained results deviate from the reported results has been scrutinized. Since diffusivity is inversely related to density, and as NPT simulation is performed, it is necessary to find out if the density of the system is equilibrated or not. The density profiles (at all the considered temperatures) after NPT runs have been analyzed. It is found that the density of the matrix decreased by approximately 5 % when compared to the initial density of the matrix. These values are reported in table 2.

The reason for this was investigated for the PIB-C320 matrix at 450 K. The initial united atom configuration is converted to all atom configurations using Material Studio *(17)*. The initial d ensity is 0.8508; the whole matrix is minimized for energy and simulated under NPT conditions for 500 ps. The averaged density of the system is rechecked and it is found to be approximately 5 % lower than the initial density. Therefore, lowering of the density of the system by Material Studio *(17)* can probably be the reason for higher value of diffusivity obtained from NPT runs. The use of NVT simulation instead of NPT, may give diffusivity value closer to the experimental/reported result as the density of the system will be kept constant all throughout the simulation.

Case study results and discussion

This case study clearly justifies the application of multilevel approach to the design of a polymer having the desired properties. Using the Macro-Meso-Scale approach, a set of basic polymer repeat units promising to have the required target property is obtained. Since only olefinic groups were included in the basis set, it is obvious that the basic polymer repeat unit structures formed from them have very low water absorptivity. This property has been validated using Van Krevelen's group contribution model *(1)*, and has been reported in table 1. Macro-Meso-Scale approach alone cannot decide the arrangement of the repeat units or the chain length, Micro-Scale approach is made used. Since it is time consuming to study all the identified repeat units from the Macro-Meso-Scale approach under Micro-scale approach and as the main intension of this work is to verify the multilevel modeling methodology, only PIB structure is used for Micro-scale approach studies.

PIB structures were investigated to see if they have low diffusivity to O_2. Though the diffusivity results obtained from simulation were two orders of magnitude when compared to the experimental results, yet the diffusivity of O_2 in PIB C-320 system is quite low. The low solubility of O_2 reported by Tsolou et al. *(10)* and low diffusivity of O_2 from the simulation result implies in the low permeability of O_2 in this designed polymer. Moreover as the molecular size of O_2 is smaller than N_2, which therefore means that this designed polymer has low permeability to N_2. In broader sense, this polymer has low permeability to air. It has also been mentioned in different literature *(12, 13, 19)* that this polymer is a very good elastomer and has low glass transition temperature too. Based on the literature sources and the obtained results from this case study, this material (PIB-C320) can be sent for the laboratory scale synthesis and also can be validated to see if it has all the required property or not.

This case study therefore clealy validates the multilevel approach for computer aided polymer design.

CONCLUSION

A multilevel approach in designing a polymer with the required properties is very promising as it decreases the time for developing a new production that is the traditional approach of synthesizing a lot of compounds in laboratory and tesiting them can be reduced to greater extent. Application of GC^+ property models for predicting polymer repeat unit structure properties in the CAMD algorithm enhances the application range of problems that can be handled. The CAMD algorithm is not a stand alone approach to the design of a new polymer based product as this approach does not give complete information on how the basic polymer repeat units are arranged or the chain length of the polymer designed. Therefore after identifying the basic repeat unit structures, using a CAMD algorithm based on Macro-Meso-Scale approach, further details on the polymer are added in terms of the arrangement of the identified polymer repeat unit to form a polymer chain by studying the conformational and structural properties of the polymer chain using atomistic simulations based on the Micro-Scale approach. This multilevel approach has been incorporated in the developed methodology.

A case study to verify/support the multilevel approach in computer aided polymer design has been perfomed. In this case study, the Macro-Meso-Micro scale approaches have been used to design a polymer that can be used for making hermetic stoppers (sealent). Polyisobutylene is one of the basic polymer repeat unit identified from the Macro-Meso-Scale approach (CAMD algorithm) and the chain length of this polymer has been determined through the Micro-Scale approach. Since Micro-scale approach is computationally intensive, it is too time consuming to rely alone on this approach. Therefore, parallel computing methods or higher-order group contribution models could be used prior to the Micro-scale approach. Thus the step wise integration of multilevel approaches helped in overcoming the disadvantages of each approaches while retaining its advantages.

REFERENCE

1. Van Krevelen, D.W. (1990). Properties of Polymers, Amsterdam: Elsevier.
2. Bicerano, J. (2002). Prediction of Polymer Properties. New York: Marcel Dekker Inc.

3. Satyanarayana, K.C., Gani, R., Abildskov, J. (2007). Polymer property modeling using grid technology for design of structured products. Fluid Phase Equilibria, 261, 58.

4. Gani, R., Harper, P.M., Hostrup, M. (2005). Automatic creation of missing groups through connectivity index for pure-component property prediction. Ind. Eng. Chem. Res., 44, 7262.

5. Manaras, C.D. (1996). Optimal computer-aided molecular design: A polymer design case study. Ind. Eng. Chem. Res.,35 (10), 3403.

6. Venkatasubramanian, V., Chan, K., Caruthers, J. M.(1994). Computer-Aided Molecular Design using Genetic Algorithms. Computers &.Chemical Engineering, 18, 9, 833.

7. Satyanarayana, K.C, Abildskov, J. , Gani, R.. Computer aided polymer design using group contribution plus property models. Computers & Chemical Engineering. (submitted 2008).

8. Harper, P.M., Gani, R. (2000). A multi-step and multi-level approach for computer aided molecular design. Computers & Chemical Engineering, 24, 2-7, 677.

9. Makrodimitri, Z.A., Dohrn, R., Economou, I.G. (2007). Atomistic simulation of poly(dimethylsiloxane): Force field development, structure and thermodynamic properties of polymer melt and solubility of n-alkanes, n-perfluoroalkanes, and noble and light gases. Macromolecules, 40, 1720.

10. Tsolou, G., Mavrantzas, V.G., Makrodimitri, Z.A., Economou, I.G., Gani, R. (2008). Atomistic simulation of the sorption of small gas molecules in polyisobutylene. Macromolecules, 41 (16), 6228.

11. Charati, S.G., Stern, S.A. (1998). Diffusion of gases in silicone polymers: Molecular dynamics simulations. Macromolecules 31, 5529.

12. Boyd, R.H., Pant, P.V. (1991). Molecular packing and diffusion in polyisobutylene. Macromolecules 24, 6325.

13. Müller-Plathe, F., Rogers, S.C., van Gunsteren, W.F. (1993). Gas sorption and transport in polyisobutylene: Equilibrium and nonequilibrium molecular dynamics simulations. J. Chem. Phys. 98 (12), 9895.

14. Marrero, J., Gani, R. (2001). Group-contribution based estimation of pure component properties. Fluid Phase Equilibria, 183-184, 183.

15. *PolyInfo* (a polymer database), National Institute for Material Science, Japan. https://polymer.nims.go.jp/PoLyInfo/

16. Harper, P.M., Gani, R., Kolar, P., Ishikawa, T. (1999). Computer-aided molecular design with combined molecular modeling and group contribution. Fluid Phase Equilibria 158-160, 337.

17. Material Studio, M.S.P., Accelrys Software Inc., San Diego, USA.

18. United States Patent 5994477 (1999), Selective hydrogenation method of living polymer having olefinic double bond.

19. Pant, P.V., Boyd, R.H. (1993). Molecular dynamics simulation of diffusion of small penetrants in polymers. Macromolecules 26, 679.

20. Karayiannis, N.C., Mavrantzas, V.G., Theodorou, D.N. (2004). Detailed atomistic simulation of the segmental dynamics and barrier properties of amorphous poly(ethylene terephthalate) and poly(ethylene isophthalate). Macromolecules, 37, 2978.

21. Karayiannis, N.C., Giannousaki, A.E., Mavrantzas, V.G. (2003). An advanced Monte Carlo method for the equilibration of model long-chain branched polymers with a well-defined molecular architecture: Detailed atomistic simulation of an h-shaped polyethylene melt. Journal of Chemical Physics, 118 (6), 2451.

22. Van Amerongen, G.J. (1950). Influence of structure of elastomers on their permeability to gases. Journal of polymer science, 5 (3), 307.

23. Kier, B., Hall, H.L. (1986). Molecular connectivity in Structure Activity Analysis. New York: John Wiley & Sons.

24. Drug Bottles Containing Natural Rubber Stoppers May Place Latex Allergic Patients At Risk for Reactions. ScienceDaily, June 15, 2001. (http://www.sciencedaily.com)

25. United States Patent 6528152 (2003), Stopper of foamed thermoplastic synthetic material.

FIGURES

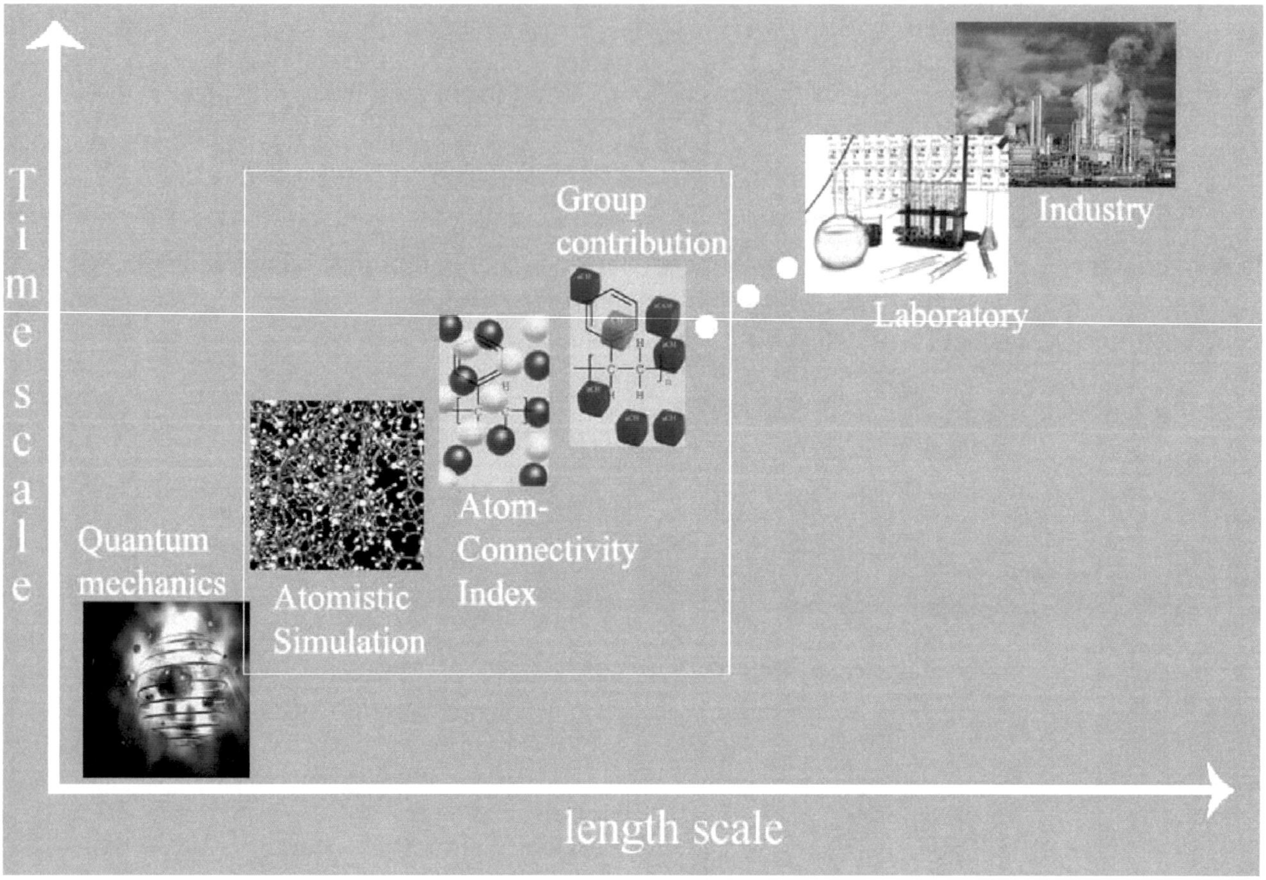

"**Figure I.** Multi scales for developing new polymer product on the basis of the required property."

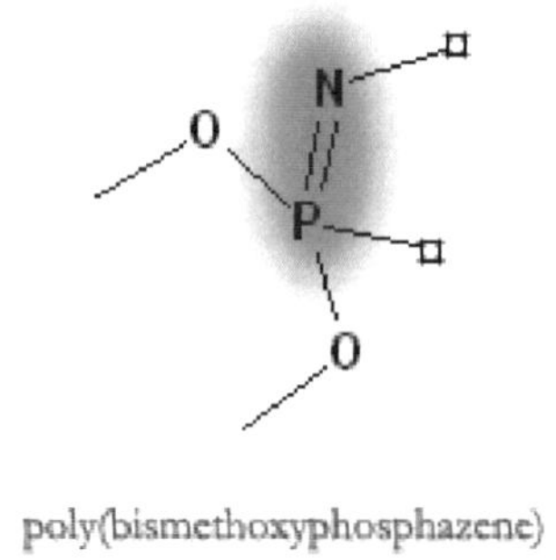

"**Figure II.** Incomplete molecular structure representation – Marrero/Gani *(14)* method does not describe the shaded groups. The small squares in the above structure denote the free ends of the polymer repeat unit structure."

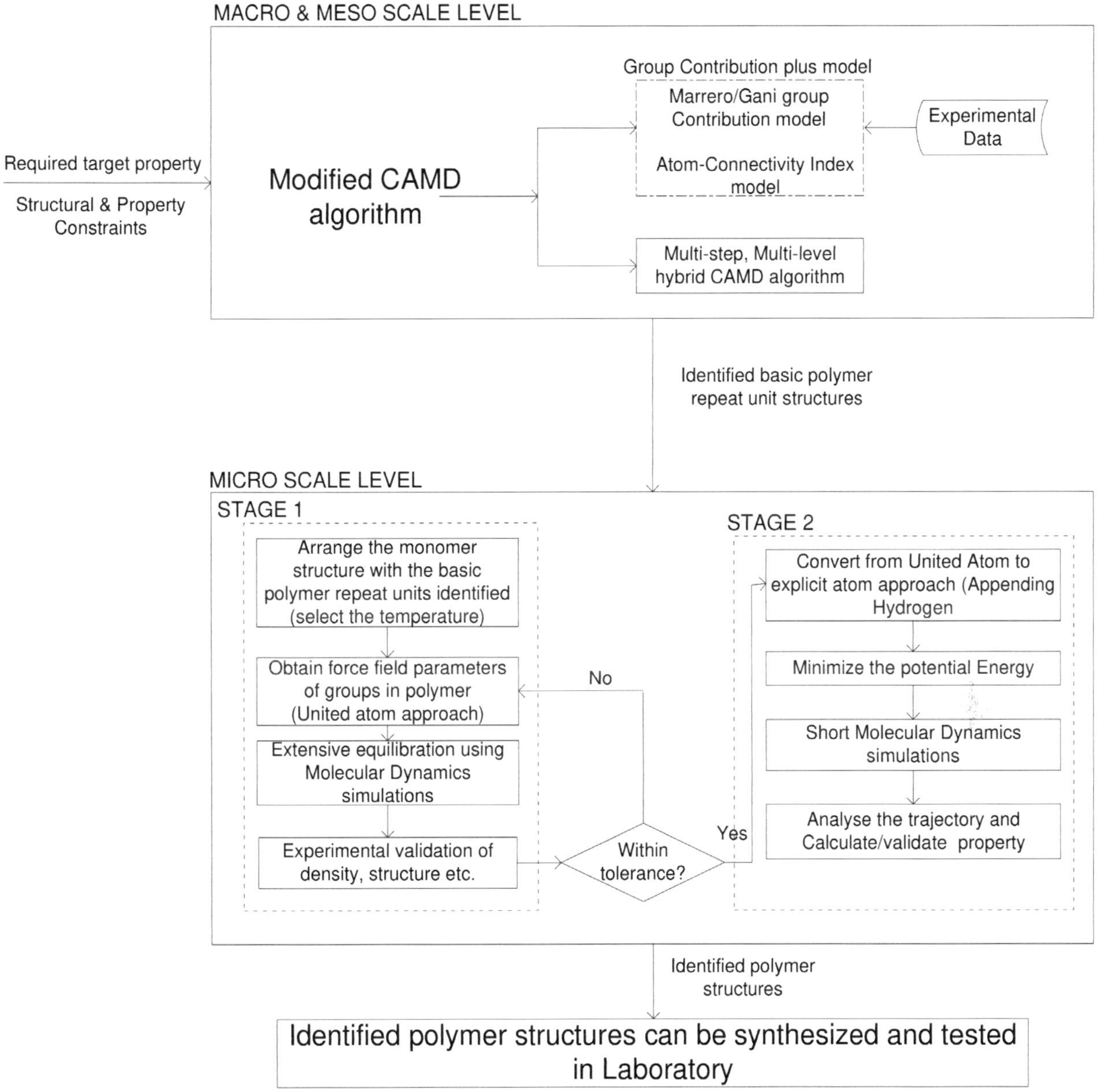

"**Figure III.** Schematic work flow of multilevel modelling for predicting polymer repeat unit structure with the required target property"

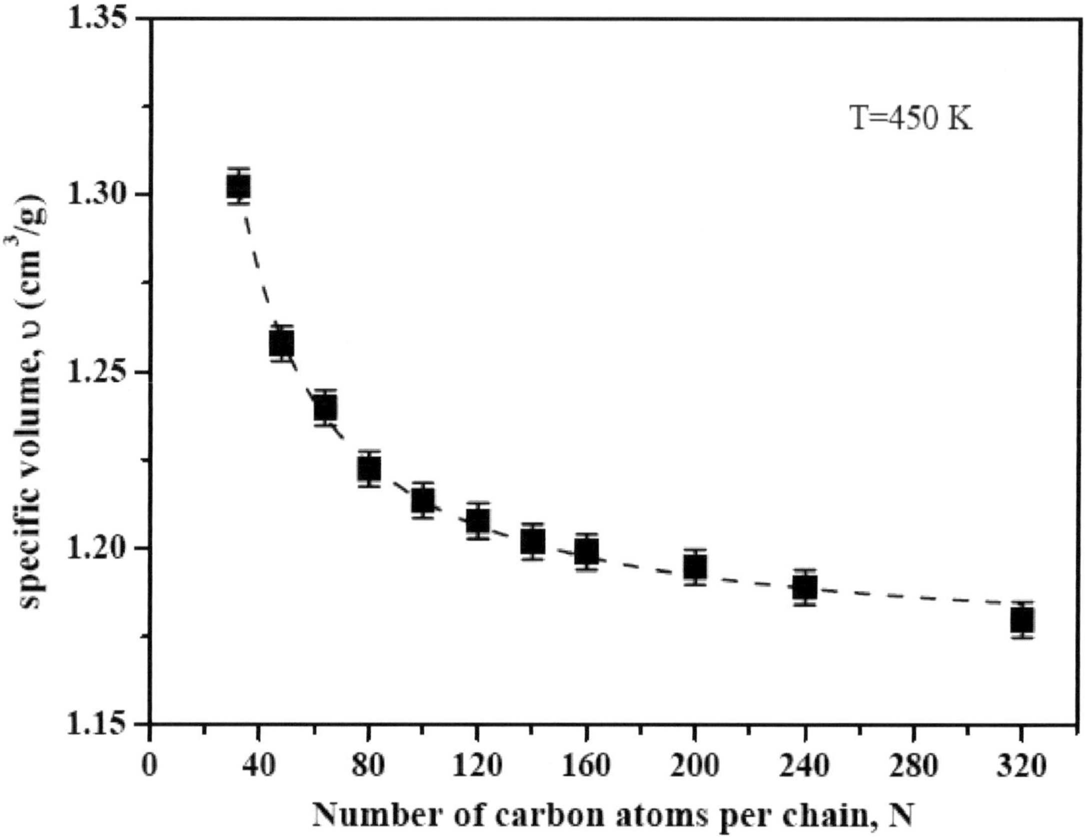

"**Figure IV.** Specific volume versus the number of carbon atoms per chain (figure reported by Tsolou et al.) *(10)*"

TABLES

Structural Constraints		
Basis set CH3, CH2, CH, C		
Number of groups	6 (Max)	2 (Min)

Property Constraints	Min	Max
Glass Transition temperature (K)		350
Dielectric Constant		2.5

Basic polymer repeat units	T_g (K)	ε	W_{abs} (g of H_2O/g of polymer)
-[-CH(CH_3)-]n-	302.51	2.48	0.00004
-[-CH_2-CH_2-]n-	168.72	2.47	0.00004
-[-CH_2-CH(CH_3)-]n-	257.91	2.37	0.00004
-[-CH_2-C(CH_3)$_2$-]n-	*342.1*	*2.23*	0.00003
-[-CH(CH(CH_3)$_2$)-]n-	303.13	2.28	0.00004
-[-CH(CH_3)-CH(CH_3)-]n-	302.51	2.26	0.00004
-[-CH((CH_2)$_2$$CH_3$)-]n-	235.61	2.25	0.00004
-[-C(CH_3)((CH_2)$_2$$CH_3$)-]n-	307.43	2.12	0.00003
-[-CH(CH(CH_3)$CH_2$$CH_3$)-]n-	275.75	2.14	0.00004
-[-CH_2-CH(CH(CH_3)$_2$)-]n-	276.24	2.17	0.00004
-[-CH(CH(CH_3)((CH_2)$_2$$CH_3$))-]n-	257.91	2.03	0.00004
-[-CH((CH_2)$_2$-CH(CH_3)$_2$)-]n-	258.32	2.05	0.00004

"**Table 1.** Identified basic polymer repeat units from CAMD algorithm"

Temperature	Initial Density	Density after NPT runs
550 K	0.798907	0.7562

500 K	0.826817	0.7781
450 K	0.8508	0.8152
350 K	0.905498	0.8658

“**Table 2.** Table showing the change in density (from initial density) after NPT runs for PIB-C320”

CHARTS

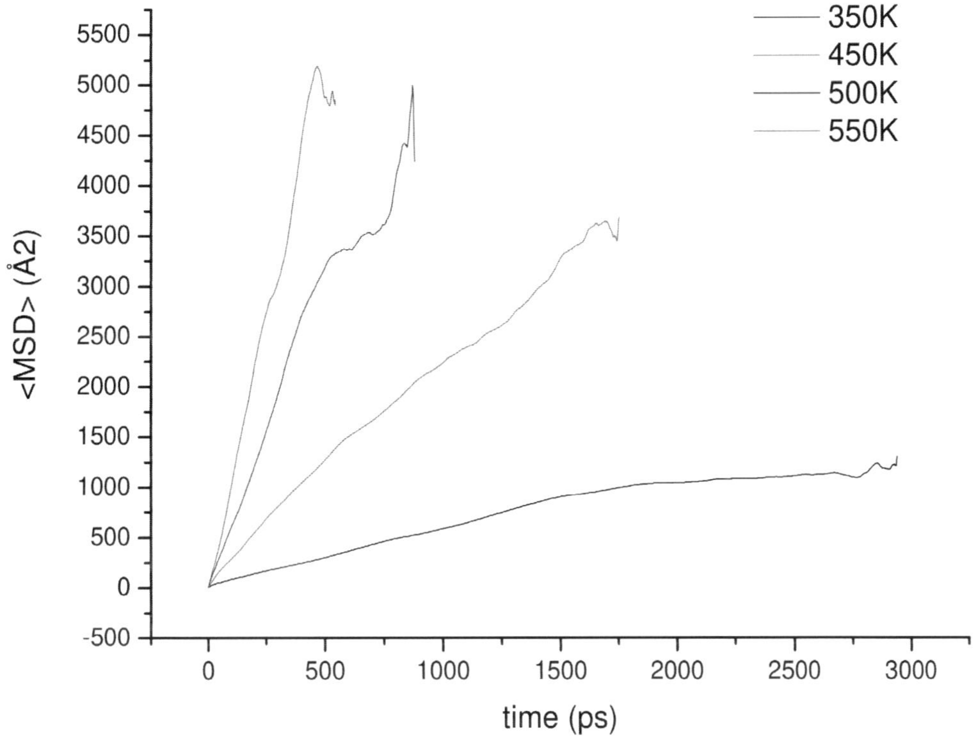

"**Chart I.** Mean square displacement of O_2 versus time in PIB-C320 matrix"

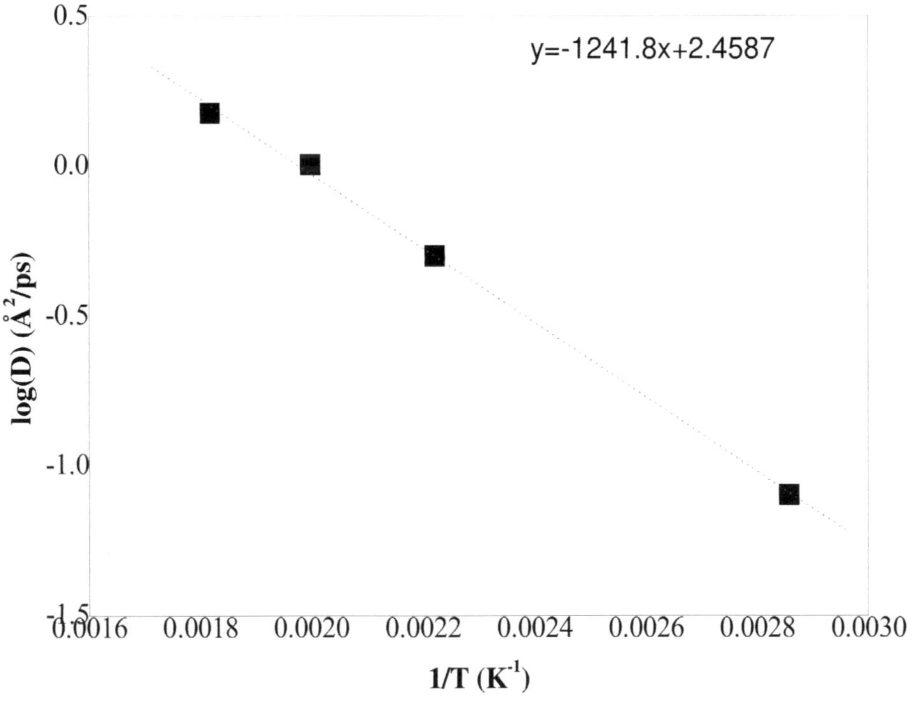

"**Chart II.** Log(D_0) versus 1/T for O_2 in PIB-C320 matrix"

717e Structured Product Design Using Combined Property Clustering and Decomposition Techniques

Charles C. Solvason, Nishanth Chemmangattuvalappil, and Mario R. Eden. Department of Chemical Engineering, Auburn University, 230 Ross Hall, Auburn University, AL 36849-5127

Recent developments in the area of integrated process and product design have shown that products can be designed in terms of their properties with out committing to any specific components a priori. However, the identification of suitable candidate molecules that satisfy a set of property targets is still a challenging issue. While current techniques make use of group contribution methods (GCM) to design molecules, there are many properties which can not be estimated by GCM. Furthermore, not all possible atomic arrangements and structures can be represented in GCM. Hence, there is a need of an efficient methodology for the design of structured molecules. One such approach to structured product design is to combine property clustering with decomposition techniques. This approach first utilizes multivariate characterization techniques to describe a set of representative samples, and then uses decomposition techniques such as principal component analysis (PCA), to find the underlying latent variable models that describe the molecule's properties. The orthogonal nature of these models allows for group-based interpretations and property predictions which can be utilized to design new molecules not found in the original set of molecules. In this work a case study of the design of an excipient for an acetaminophen tablet is presented.

717f Design of Secondary Refrigerants. A Combined Optimization-Enumeration Approach

Apurva Samudra and Nick Sahinidis. Chemical Engineering, Carnegie Mellon University, 5000 Forbes Avenue, Pittsburgh, PA 15213

Computer-aided molecular design has emerged as a powerful technique to identify promising compounds that meet the predefined property targets. These techniques have been employed in various areas, including the design of solvents, refrigerants, and polymers.

This paper will present a mixed-integer non-linear program (MINLP) for designing a secondary refrigerant. Secondary refrigeration loops have recently been recognized as a potential solution to reduce the environmental impact of retail food refrigeration. The MINLP model adopted here includes a property prediction model (Nanda, 2001) coupled with molecular structure feasibility constraints (Sahinidis et al., 2003). A large number of basic building groups is considered and no limits on the type of the molecule are imposed. To cope with the numerical challenges associated with nonlinearities of this model, a direct enumeration of the search space is used to solve the model for restricted molecular sizes, while a branch-and-bound algorithm is used for larger designs. Several novel solutions are obtained and will be presented and analyzed in detail. We also explore the possibility of using accurate property models (Marrero and Gani, 2001) along with the possibility of extending these models to include higher-level properties which are structure-dependent for larger molecules.

Bibliography:

Marrero J. and R. Gani, A group contribution based estimation of pure component properties, Fluid Phase Equilibria, 183, 183-208, 2001.

Nanda, G., Design of efficient secondary refrigerants, M.S. thesis, University of Illinois at Urbana-Champaign, 2001.

Sahinidis, N. V., M. Tawarmalani, and M. Yu, Design of alternative refrigerants via global optimization, AIChE Journal, 49(7), 1761-1775, 2003.

717g Lattice Monte Carlo Modeling and Analysis of Microstructure and Coating Quality Correlation for Optimal Paint Design

Jie Xiao and **Yinlun Huang**. Department of Chemical Engineering and Materials Science, Wayne State University, 5050 Anthony Wayne Drive, Detroit, MI 48202

Paint is designed to offer various chemical and physical properties for surface protection, styling, and appearance. Nevertheless, the anticipated coating quality is frequently unsatisfactory, which is often attributed to paint formulation design. As new demands on coating performance continuously emerge, paint formulation design becomes even more sophisticated. This has triggered computational paint design as it can provide great freedom and control over the investigation of paint formulation through any number of in silico experiments under any application conditions.

In this paper, we introduce a comprehensive microscale modeling and analysis methodology for the design of paint formulation and the prediction of paint-based surface coating properties. By this methodology, which is resorted to lattice Monte Carlo (LMC) modeling and simulation, any paint formulation can be evaluated under any real industrial curing condition, and the coating performance in terms of solvent resistance and inter-coat adhesion can be thoroughly investigated. This can enable an establishment of various types of correlations among paint formulation, curing condition, coating microstructure and coating properties, which are critical for paint designers to optimally develop paint formulation. The introduced methodology is used to study the construction and analysis of polymer networks, where the distributions of polymer molecular weights and functional groups on polymer chains are all taken into account. Different from the known LMC simulation techniques, a unique approach is introduced to impose a curing condition on the network formation process, which enables the investigation of curing dynamics along both real curing time and MC steps. Furthermore, a new network analysis method is described for evaluating a key coating quality indicator, EECD (elastically effective crosslink density) by utilizing the information contained in a 3D polymer network structure. This can help greatly bridge the gap between the investigation of paint microstructures and the research on coating macroscopic quality. A comprehensive study on optimal design of acrylic-melamine-based resin will demonstrate the efficacy of the introduced methodology.

AUTHOR INDEX

AUTHOR INDEX

AUTHOR INDEX